Dictionary
of
Inventions
and
Discoveries

DICTIONARY
OF
INVENTIONS
AND
DISCOVERIES

E. F. Carter

PHILOSOPHICAL LIBRARY
NEW YORK

Published, 1967, by Philosophical Library Inc.,
15 East 40th Street, New York 16, N.Y.

Printed in Great Britain for Philosophical Library
by The Garden City Press Limited
Letchworth, Hertfordshire

Introduction

THE purpose of this work is to set out in alphabetical order the scientific, technological, and kindred inventions, discoveries and developments which have been made since earliest times, indicating to whom each is attributable, together with the year in which they were made or introduced, if known.

This knowledge has been gathered from many thousands of literary and factual sources, and represents the culmination of many years of intensive research, inquiry and travel. But a work of such coverage must, at very best, leave much to be desired. Any attempt to cover entirely the vast field of human endeavour down the ages must necessarily fall short of completeness.

In this the serious student is invited to explore more fully the history of science, technology and engineering; but he must be prepared to accept that much early historical evidence still remains buried in original documents and periodicals in many languages. Here, then, is a wide-open field for his patient research and evaluation of the abundant material awaiting his efforts; and if this book helps him on his way, its purpose will have been fulfilled.

ERNEST F. CARTER
WARLINGHAM, SURREY
JANUARY 1966

ABBREVIATIONS

Appl.	Applied	Bel	Belgian
Att.	Attributed to	Ch	Chinese
Const.	Constructed	Aus	Austrian
Dem.	Demonstrated	Fr	French
Des.	Designed	Ger	German
Desc.	Described	Hol	Dutch
Dev.	Developed	Hun	Hungarian
Disc.	Discovered, Discovers	It	Italian
Est.	Established	Jap	Japan
Form.	Formulated	Jugo-slav	Jugoslavian
Illus.	Illustrated		
Imp.	Improved	Port	Portuguese
Inst.	Instituted	Nor	Norwegian
Intro.	Introduced	Dan	Danish
Inv.	Invented	Rus	Russian
Isol.	Isolated	Scot	Scottish
Leg.	Legendary	Sp	Spanish
Obs.	Observed	Swed	Sweden
Obt.	Obtained	Swit	Switzerland
Pat.	Patent, patented	U.S.	United States of
Prep.	Prepared		America
Prod.	Produced		
Prop.	Propounded	Per.	Persia, Persian
Pub.	Published	A.G.	Ancient Greek
b.	Born	A.E.	Ancient Egypt.
d.	Died	Asy.	Assyrian
n.d.	No date	Gr.	Greece, Greek
		A.R.	Ancient Roman
Braz	Brazilian	Phon.	Phoenician

A

ABBÉ, Ernst (Ger) (1840–1905). Friend of Carl Zeiss. *See* Glass, Optical.

Aberration *See* Light.

Absorbent Dressing (surgery) *Inv.* Dr. Matthias, Mayor of Lausanne, Switzerland.

Absorbiometer 1855. *Inv.* Robert Wilhelm Bunsen (Ger) (1811–99).

Acarus (Itch insect) *Disc.* Ebn Zoar (Avenzoar) (Arab) (1170–1262).

Accelerator, Linear 1932. *Inv.* with Ernest Walton by John Cockroft.

Accelerometer 1783. (Machine for proving acceleration) *inv.* George Attwood (1746–1807). 1909. Modern A. *inv.* H. E. Wimperis and G. K. B. Elphinstone.

Accents (language) 264 B.C. *Intro.* into Greek language by Aristophanes of Byzantium. 1610. First used in French language.

Accordion *Inv.* 1822. Buschmann of Berlin. (Other sources cite *inv.* 1829 by Damian of Vienna.) *Intro.* G.B. 1829 from Vienna.

ACCUM, Fredrich (1769–1838). (Ger). Credited with *intro.* of gas lighting. *See* Gas, coal.

Accumulator, Electric 1803. First attempt to produce a storage battery made by Johann Wilhelm Ritter (1776–1810) (Ger). 1859. Lead-celled Accumulator *inv.* Raimond-Louis-Gaston Planté (1834–9) (Fr). 1881. Formed and coated Accumulator grid *inv.* C. A. Faure (Fr). 1881. Sir Joseph Wilson Swan *pat.* lead-grid Accumulator. 1888. Modern type Accumulator grid *inv.* Karl Erich Correns (1864–1933) (Ger). 1900 c. Thomas Alva Edison (1847–1931) *inv.* nickel-alkaline A.

Accumulator, Hydraulic, *Inv.* Sir William Armstrong.

Acetaldehyde 1881. *Disc.* Kutscherov (Rus), by passing acetylene gas through sulphuric acid in presence of mercury as a catalyst.

Acetylene (gas) 1836. *Disc.* Edmund Davy, obtained from potassium carbide. 1896. G. Claude and A. Hess first solve Acetylene in acetone. c. 1870. *Prod.* from carbon and hydrogen by electric arc by P. E. M. Berthelot (Fr). 1924. Acetylene first inhaled as an anæsthetic. (*See also* Synthetic Rubber, P.V.C. and Vinyl Acetate.)

ACHARD, Franz C. (1753–

1821) (Ger). 1801. Built first sugar-beet factory.

Achromatism 1729. Chester Moor Hall *disc.* how to correct chromatic aberration by combining lenses of flint and crown glass. 1755. Achromatic lenses *dem.* by Samuel Klingensteirna (Swed). 1758. Hall's *disc. re-disc.* by John Dolland.

Acoustics of Buildings 1809. First theory advanced by Sir George Cayley (1773–1857).

Acre (measurement) 1305. English statute Acre defined precisely as 4,840 square yards.

Acromegaly (medicine) 1885. First *disc.* by Pierre Marie (Fr).

Acrostic *Inv.* A.D. 328 by Porphyrius Optaliamus (Malchus) of Tyre. (*c.* 233–404.)

Actinium (element) 1899. *Disc.* M. Debierne (Fr) in pitchblende.

Actinometer 1825 *Des.* Sir John F. Herschel (1792–1871) (Ger). 1856. *Inv.* Robert Wilhelm Bunsen (1811–99) (Ger) and Sir Henry Roscoe (1833–1915).

ADAMS, John Couch (1819–92). 1846. *Disc.* planet Neptune.

Addison's Disease (skin). First diagnosed by Doctor Thomas Addison, of Guy's Hospital, London.

Adrenaline, or **Epinephrine** 1901. Synthesized by Joachi Takamine (1854–1922). Also independently by Thomas Bell Aldrich. (First *isol.* of a hormone.)

Advertisement 1652. First advertisement appeared in a Parliamentary paper.

Aelopile (jet steam-engine) First recorded by Heron of Alexandria (*c.* 285–222 B.C.). *See also* Turbine, Steam and Steam Engine.

Aeolian Harp Known and possibly *inv.* by Duncan, Archbishop of Canterbury in 10th cent. Generally ascribed to Athanasius Kircher (1601–80) (Ger).

Aerial, Directional 1905. *Inv.* Gugliemo Marconi (It) (1874–1937).

AGRICOLA, Georgius (Georg Bauer) (1494–1555) (Ger). Author of *De Re Metallica, pub.* in Basle, 1556. *See* Pumps, Gears, and General engineering.

Agriculture, application of steam to 1684. The brothers William, Andrew and John Parham, and Thomas Dornaxy pat. idea for steam machine for agricultural work. 1767. Francis Moore *pat.* machine of wood and metal worked by fire and water or air for agricultural purposes. *See also* Traction engine, Steam ploughing engines, Steam pumps, etc.

Aileron (aircraft) 1868. *Inv.* Matthew Piers Watt Boulton. (*Pat.* No. 392.)

Air *See* Atmosphere, composition of.

Air Compressor Peter Brotherhood (1839–1902). *Pat.* three-stage dry air compressor.

Aircraft (including heavier and lighter than air machines, kites and rockets; but not including space-craft (q.v.)) 400 B.C. Archytas of Tarentum reported to have made a steam-jet propelled pigeon. 4th cent. B.C. Kite *inv.* in China. 4th cent. A.D. Probable first conception of airscrew made in China. *c.* 1020. Oliver of Malmesbury attempts to fly. *c.* 1250. Roger Bacon speculates about flight. 1379. First mention of rockets made in Europe by Muratori. *c.* 1500. Designs for a helicopter, aeroengine, parachute, retractable under-carriage and two types of powered projectiles made by Leonardo da Vinci. 1503. G. B. Danti attempts to fly at Perugia. 1507. John Damian (It) attempts to fly at Stirling, Scotland. 1560. First mention of Kite in Europe by Schmidlap. 1589. First *desc.* of kite by della Porta. *c.* 1620. Veranzio *pub. desc.* of a parachute. 1634. John Bate *pub.* first drawing in Europe of a kite. 1631. Desmarets *pub. illus.* of parachute. *c.* 1655. Model ornithopter tested by Robert Hooke. 1660. Allard attempts to fly before Louis XIV at St. Germain. 1678. Besnier attempts to fly at Sablé. 1680. Borelli *pub.* report of a demonstration of the inadequacy of muscle-powered flight. 1709. Gusmão *des.* the "Passarola." 1714–15. Swedenborg *des.* an idea for an aeroplane. *c.* 1742. De Bacqueville attempts to fly across the Seine at Paris. 1746. Robins first uses the whirling arm. 1768. Paucton *des.* his "pterophore" machine. 1772. Desforges builds his "voiture volante." 1781. Meerwein builds an ornithopter-cum-glider. 1783 (Nov. 21). First human aerial voyage made by de Rozier and d'Arlandes in a Montgolfier hot-air balloon at Paris. 1783

(Dec. 1). First voyage in hydrogen balloon made by Charles and Robert. 1784. First successful test in Europe of a model helicopter by Launoy and Bienvenu, using a bowstring motor. 1784. Valet tests first full-sized airscrew on a river boat. 1784. Blanchard makes first attempt to propel a balloon by airscrew. 1784. First aerial voyage in Britain by Lunardi in a hydrogen balloon. 1784. Meusnier *des.* dirigible with screw propeller. 1784. Gérard made first suggestion to propel aircraft by rockets. 1784. Rénaux and Gérard *des.* ornithopters. 1784. Aries attempts to fly at Embrun. 1785. Blanchard and Jeffries make first crossing of the English Channel by air in a balloon. 1785. Death of Romain and de Rozier—the first ballooning fatalities. 1792. Sir John Cayley flies his model helicopter. 1794. First military reconnaissance from a balloon made at Maubeuge. 1797. First human parachute descent made by Garnerin. 1799. First *des.* of modern configuration of aeroplane made by Cayley. 1804. Cayley tests aerofoils on a whirling arm. 1805. Congreve *des.* modern type rocket missiles. 1805–9. Cayley flies first unmanned, full-sized glider. 1806–17. Dagen attempts to fly. 1807. Cayley tests modified gunpowder engine as power-source. 1809. Cayley *des.* solids of least air resistance. 1810. Cayley *pub.* papers on heavier-than-air flight. 1811. Beblinger attempts to fly at Ulm. 1818. De Lambertye *des.*

an ornithopter and a helicopter. *c.* 1825. Pocock flies a man-lifting kite at Bristol. 1827. Pocock travels in a kite-drawn carriage. 1828. Vittorio Sarti *des.* a contra-rotating device. *c.* 1830. Artingsall tests model ornithopter. 1837. Cocking killed in a parachute drop. 1842. W. H. Phillips helicopter flies with steam-jets on rotor tips. 1842. W. S. Henson, of Chard, Somerset *des.* his "Aerial Steam Carriage." 1843. Engineer Bourne flies model helicopter. 1843. Dr. William Miller *pub. desc.* of his ornithopter. 1843. Cayley *des.* a covertiplane. 1847. W. S. Henson tests his first powered model monoplane. 1848. John Stringfellow (1799–1883) tests powered aeroplane. 1849. Austrian Montgolfier pilotless balloon makes first bombing raid in history—against Venice. 1852 (Sept. 24). First flight by Henri Giffard's full-sized, manned dirigible—the first full-sized aircraft to be powered by an aero engine. 1852–53. First man-carrying flights by a heavier-than-air machine—one of Cayley's gliders. 1852. Michel Loup *des.* aeroplane. 1854. François Letur killed on a parachute glider. 1856. Carlingford *des.* an aeroplane. 1856–68. J. M. Le Bris, a French sea-captain tests full-sized gliders. 1856–96. L. P. Mouillard tests full-size gliders. 1857–58. Felix du Temple (1823–90) made first successful powered flight in a modified aeroplane. 1858. First aerial photograph made from captive balloon "Nadir," over Paris.

1858. Pierre Jullien made twin-propellered model aeroplane with a twisted-rubber motor which flew 39 ft. 1858–59 F. H. Wenham (1824–1908) makes multi-planed glider. 1859. E. S. Cordner, an Irish priest flies a man-carrying kite for rescue work. 1860. Smythies *des.* ornithopter-cum-fixed-wing aeroplane. 1863. Gabriel de la Landell *des.* a fantastic helicopter. 1863. Vicomte de Ponton d'Amécourt tests steam-driven helicopter model. 1863. Jules Verne *pub.* his story "Five Weeks in a Balloon." 1865. de Louvrié, a French engineer *des.* earliest known jet-propelled heavier-than-air aeroplane. 1867. Butler and Edwards *des.* and *pat.* delta-wing jet-powered aeroplane. 1868. John Stringfellow tests first powered triplane, with modern aileron system *inv.* by Boulton but not *pat.* 1868. J. M. Kaufmann, of Glasgow tests steam-driven ornithopter with a modified fixed wing. 1868. J. J. Bourcart tests his ornithopter at Guebwiller. 1870. Alphonse Pénard (1850–80) *intro.* model with twisted-rubber power. 1870. G. Trouvé flies gunpowder-driven model aeroplane. 1870s. Danjard flies tandem-winged glider. 1871. Pénard *des.* his model "Plano-phore" and flies it. 1871. First wind-tunnel *inv.* by F. H. Wenham and John Browning. 1871. John Goodman Household, of South Africa flew $\frac{3}{4}$ mile in a glider of his own *des.* Haenlein flies first full-sized aircraft fitted with an internal-combustion engine—a dirigible. *c.* 1874. Félix Du Temple flies first powered man-carrying aeroplane at Brest. 1875. Thomas Moy tests large tandem-wing "Aerial Steamer." 1876. Pénaud *pat.* full-sized *des.* as an amphibian. 1877. Mélikoff *des.* gas-turbine model helicopter. 1877. Enrico Forlanini (It) *des.* steam-driven helicopter which flew. 1879. Victor Tatin *des.* a compressed-air-driven model monoplane. 1878. Russians claim that A. Mozhaiski flew. 1878. Biot tests full-sized glider. 1878. Brearley *des.* model "undulator" which flew. Dandrieux *des.* model "butterfly" helicopter which flew. 1880. Linfield, of Margate *des.* a 240 lb. machine driven by a 9-bladed airscrew 7 ft. in diameter, revolving at 112 r.p.m. 1883. A. Tissandier *des.* first electrically-driven aircraft—a full-sized dirigible. 1884. A. Mozhaiski (Rus) tests powered man-carrying aeroplane. 1884. Renard and Krebs fly in first practicable dirigible. 1884. Phillips *pat.* cambered wings. 1887. Hargraves *inv.* rotary engine. 1888. First full-sized aircraft to be powered in flight by a petrol engine—Wölfert's dirigible. 1890 (Oct. 9). Clement Ader's "Eole" the first full-sized aircraft to leave the ground under its own power. 1891. Prof. Langley tests his first steam-powered, tandem-wing model. 1891–96. Otto Lilienthal pilots many successful glider flights. 1893. Horatio Phillips tests large multi-wing model aeroplane. 1893. Hargreaves *inv.* box-

kite. 1894. Baden-Powell *des.* first practical man-lifting kites. 1894. Sir Hiram Maxim (1840–1916) tests full-sized aeroplane of his own *des.* 1895. Lilienthal flies first piloted biplane glider. 1895. Percy S. Pilcher (1866–99), the first British aviator, makes first successful piloted glider flight on Clydebank in "The Bat" (his glider "Hawk" was flown at Eynsford, Kent, the following year). 1896–97. Samuel Pierpoint Langley, Victor Tatin, and Charles Richet fly their model steam aeroplanes. 1897. Wilbur Wright (U.S.) *inv.* warping-wing system. 1900. Count Zeppelin's first airship flies. 1901. The Wright Bros. fly their first glider. 1901. Langley flies his first petrol-driven model aeroplane. Kress's first full-sized petrol-driven aeroplane destroyed while taxying. 1901. Alberto Santos Dumont (Braz) flies his No. 6 dirigible round the Eiffel Tower. 1901. Col. Cody *pat.* man-lifting kite. 1903. Lebaudy Bros. fly first fully practicable dirigible. 1903. Karl Jatho (Ger) tests aeroplane. 1903. Prof. Langley's man-carrying "aerodrome" crashes on take-off. 1903 (Dec. 17). First powered, sustained and controlled aeroplane flight in history made by Wilbur and Orville Wright (U.S.). 1904. Ailerons and elevons first used by Robert Esnault-Pelterie (Fr). 1904–5. J. H. C. Elhammer (Dan) tests tethered aeroplane on Lindholm Island, and made a powered leap in 1906. 1905. Gabriel Voisin, Ernest Archdeacon and Louis Bleriot test float-gliders and open the first aeroplane factory in the world at Billancourt, Paris. 1905. Capt. F. Ferber *intro.* a stable, powered tractor biplane. 1906. First "Antoinette" aero engine of 24 h.p. in service. 1906. Trajan Vuia (Hun) tests tractor monoplane. 1906. Santos Dumont makes first official powered flight in Europe. 1907. Phillips makes first tentative flight in Britain. 1907. J. W. Dunne tested first swept-wing tailless biplane at Blair Atholl, Scotland. 1907. Col. Cody flies unmanned engine-driven kite. 1907. Bleriot tests first cantilever monoplanes "Canard" and "Libellule." 1908 (Sept. 29). Bros. Breguet *des.* 4-rotored, helicopter driven by a 50 h.p. Antoinette engine, which carried a man while flying tethered at Douai. 1908 (Nov. 13). Paul Cornu *des.* helicopter with two rotors driven by a 24 h.p. engine which carried a man in free flight at Lisieux. 1908. Flap-type ailerons first used by Louis Bleriot on his "Bleriot VIII" machine, which made the first monoplane flight— 5 minutes. 1908. R. Lorin *pub. des.* for jet aeroplanes. 1908 (Oct. 3). Col. Cody makes first powered flight in Britain. 1907-15. Long-range aerial bombardment of Berlin proposed by Réne Lérin (Fr) with a "torpille ærienne" which was to be catapult-launched, powered with a pulse-duct motor, radio and gyro controlled, and a 400 lb. explosive charge. It was turned down by the French Government. 1908. A. Goupy tests his triplane. 1909. 50 h.p. 7-cylinder Gnome rotary engine went into service. 1910. Henri Coanda (Roum-Fr) *des.* unsuccessful jet-propelled aeroplane with a ducted "turbo-propulseur." 1910. A. Etévé *inv.* the airspeed indicator. 1910. Hugo Junkers (Ger) *pat.* deep cantilevered wing monoplane—the "J-1," which first flew in 1915. 1910. Geoffrey de Haviland builds two aeroplanes. 1911. Wieneziers (Ger) *inv.* first retractable under-carriage. 1911. Esnault-Pelterie (Fr) *inv.* oleo under-carriage. 1912. Ruchounet (Fr) *inv.* monocoque *const.* applied by Bechereau in his Deperdussin monoplane. 1912. First all-metal aeroplane—the "Tubavion" flies. 1912. First enclosed cabin aeroplanes— the Avro mono and biplanes, built. 1912. Reissner *inv.* corrugated aluminium wings. 1913. Dunne flew his tail-less "No. 8" from Eastchurch to Paris. 1919. Handley Page *pat.* slotted wing. 1919. Rohrbach *pat.* stressed-skin metal *constr.* 1920. Poulain leaps 40 ft. on winged pedal-cycle. 1920. First modern-type retractable under-carriage fitted to U.S. Dayton-Wright aeroplane. 1923. First refuelling in flight accomplished. *c.* 1924. Experiments commenced in Germany on what was later known as the "V-1" flying-bomb; "Ascania" (code name) working on tail and control *des.,* "Argus" on the pulse-motor, and Lusser, of Messrs. Fieseler, of Cassel on the body. 1924. Variable-pitch propeller *inv.* by Dr. H. S. Hely-Shaw and T. E. Beacham. 1925. First De Haviland "Moth" flew. 1926. Prof. G. T. Hill flies his tail-less "Pterodactyl," which was the subject of intense research in Germany by A. Lippisch and H. Koehl. 1926. Godard "flies" first liquid-fuel rocket. 1926–28. A. A. Griffith first proposed and experimented with a gas-turbine/ propellor. 1928. F. Stamer, from the Wasserkupp flies first rocket-propelled glider. 1929. Fritz von Opel made first successful jet aeroplane flight in history on a cradle-launched rocket-glider, flying for 10 minutes and reached a speed of 100 m.p.h. 1929. Sperry perfects artificial horizon and directional gyro. 1930. Frank Whittle took out his first turbo-jet *pats.* 1931. Lippisch flies delta-wing aeroplane. 1932. First practical variable-pitch propeller *intro.* 1934. First practical constant-speed propeller *intro.* 1935. R. J. Mitchell (1895–1937) *des.* and made a mock-up "Spitfire." 1936 (Mar. 5). "Spitfire" first flies and 450 ordered by British Air Ministry. 1936. Focke and Dr. Angelis *des.* first practical helicopter. 1937. First pressurized-cabin aeroplane—the Lockheed XC-35, flies. 1939. First turbo-jet aeroplane—the Heinkel 178, flies; powered by Dr. H. von Chain's centrifugal-flow, turbo-jet engine. 1940. Caproni-Campini quasi-jet aeroplane first flies. 1941. First British turbo-jet—the Gloster E28/39, flies with Whittle engine. 1942. First U.S. turbo-jet

—the Bell XP-59a flies with Whittle-type engine. 1942 (Dec). German "V-1" launched to glide from "Kondor" aeroplane. 1942 (Dec. 24). First powered launch of a "V-1." 1943. "D.H.-100" "Vampire" flies with centrifugal-flow gas-turbine *des.* by Maj. F. B. Halford. 1943. First jet-rotored helicopter, *des.* by Doblhoff, flies. 1944. First practical rocket-propelled aeroplane—the Messerschmitt "Me-163," flies. 1944 (June 12). First "V-1" flying-bombs fall on England (10 were fired, out of which only 6 reached England, 4 crashing at launch and blowing-up). 1944. German long-range rocket-bombs ("A-4") in action against England. 1945. First turbo-prop aeroplane—a Rolls-Royce Trent-engined Gloster Meteor, flies. 1947. Largest aeroplane ever built—the Hughes "Hercules," flew, but was unsuccessful. (It was a flying-boat with a 320 ft. wing-span, powered by $8 \times 3,000$ h.p. engines and was to have carried 700 persons.) 1948. First turbopropeller air liner—Vickers Viscount, flies. 1948. First jet-propelled delta-wing—Convair "XF-92A," flies. 1949. First D. H. "Comet" flies. 1949. Leduc "010" ram-jet flies. 1950. First ram jet helicopter—the Hillier "Hornet," flies. 1954. First wingless V.T.O.L. (vertical take-off and land) aircraft flies with Rolls-Royce engines. 1955. First convertiplane makes successful translation from vertical to horizontal flight (the McDonnell XV-1). 1955. First flying-platform helicopter flies. 1955. First aeroplane with inflatable fabric wings (the M.L. "Utility") flies. 1956. Peter Twiss first pilot to exceed 1,000 m.p.h. in a Fairey Delta II research aeroplane. 1965 (May 1). Col. Robert L. Stevens and Lieut.-Col. Daniel André average 2,062 m.p.h. in a U.S. Lockheed "YF-12A" aeroplane, at Edwards Air Force Base, California. 1965 (later). John Mackay reached a speed of 3,614 m.p.h. in an "X-15A" rocket-plane at Edwards Air Force Base. (*See also* under individual entries—*e.g.* Gyro-compass; and under Balloon, Dirigible, Engine, Jet, etc.)

Air-gun 13th cent. Pea-shooters depicted on two illuminated manuscripts. *c.* 1320. French manuscript shows figure aiming a blow-gun at a rabbit. 1425. Blow-gun known as *cerbottana* in Italy. (Cf. Arabic *zabatana* and Malayan *sumpitan*. 1495. Air-gun *inv.* Leonardo di Vinci. 1605. Martin Bourgoise, of Lisieux, France, *inv.* air-gun. 1607. Bartolomeo Crescentio (It) *des.* air-gun equipped with a powerful spring. 1644. Martin Mersenne (1586–1648) (Fr) mentions air-gun. 1648. John Wilkins (1614–72) mentions "wind-gun." 1656. Air-gun *inv.* Guter, of Nuremburg.

ALBERTUS Magnus (Albrecht of Cologne) (1206–80) Pioneer of modern biology, embriology and botany.

Alchemy A.D. 410. First book on by Zosimus of Panopolis, Greece.

Alcohol 1100. First redistilled at Salerno, Italy. 1320. First large-scale production at Moden, Italy. 12th cent. First obtained by Abucasis. 1830. Still *inv.* Aeneas Coffey for production of high purity alcohol. for commercial purposes. *See also* Distillation and Still.

Alcohol, Amyl 1849. *Isol.* by Prof. Edward Frankland.

Alcoholometer *Inv.* Gay-Lussac (Fr) (1778–1850).

Aldehyde 1782. *Disc.* Carl Wilhelm Scheele (1742–86) (Ger).

Algebra *c.* A.D. 250. Diophantus of Alexandria said to have *inv.* and written first book on algebra. 9th cent. Algebra cultivated by Arabs. 1220. Leonardo Bonnacio, of Pisa developed Algabra in Italy. 1494. Luca Paciolo (It) published first printed book in Europe—on Algebra. 1522–26. Algabraic signs *intro.* Christophe Rudolph and Michael Stifelius, of Nuremburg. 1590. Francis Vieta *imp.* Algebraic signs, which then came into general use. 1637. René Descartes (1596–1650) (Fr) first applied algebra to geometry.

Algebraic Signs + and — signs first used by Johann Widmann (Ger). Michael Stifel (Ger) first used signs $\sqrt{}$, $\sqrt[3]{}$ and $\sqrt[4]{}$ for square, cube, and 4th roots. 1634. Simon Stevin (Stevinius) (1548–1620), of Bruges, first used sign = as equals. 1574–1660 William Aughtred *inv.* signs ×, for multiplication, +, for addition, and ::, for proportional to. (Only two of his original symbols now in use.) Thomas

Harriot (1560–1621) *inv.* signs > (greater than), and < (less than); which were used from 1631. 1603. Mantissa *inv.* by John Wallis (1616–1703).

Alizarin (red dye) 1831. *Disc.* Robiquet and Colin. 1866. Obtained by Sir William Perkin (1828–1907). 1869. Independently *obt.* from anthracine by Liebermann and Græbe (Ger). (1865 *Disc.* by Heinrich Caro.)

Alkalimeter 1816. *Inv.* Dr. Andrew Ure, of Glasgow. Type *inv.* Gay-Lussac (Fr) (1778–1850).

Allotropy (chemical) 1841. Jöns Jacob Berzelius (1779–1848) (Swed) first observed phenomena by converting charcoal into graphite. 1844. Jean-Baptiste-Joseph Foucault (1819–68) (Fr) and Hippolyte Fizeau (1819–96) (Fr) clarified the theory of allotropy. 1893. Henri Moisson (1852–1907) (Fr) first *prod.* diamonds from carbon. As did Ruff in 1917.

Almanac 1150. Solomon Jarchus published manuscript almanac. 1457. Astronomer Purbach published first printed almanac, which continued yearly for 30 years. 1497. First almanac appeared in England, translated from the French. 1767. Dr. Maskelyne published the first Nautical Almanac in England. 1680. Francis Moore's first almanac published. ("Old Moore's Almanack" of later years, then known as "Vox Stellarum.") 1834. First "modern" Nautical Almanac published by British Admiralty.

Alpaca (textiles) 1836. *Intro.* England by Earl Derby, the first weaving made at Bradford, Yorkshire by Sir Titus Salt. 1850. Alpaca first used for umbrellas by Sangster. 1852. Full-scale alpaca weaveries *est.* by Sir Titus Salt at Saltaire, near Shipley, Yorkshire.

Alternator, Electric 1884. *Inv.* Nicolo Tesla (also Bradley and Haselwander).

Alum *c.* 1300. *Disc.* at Rocca, Syria. 1470. *Disc.* in Tuscany. 1757. *Disc.* in Ireland. 1790. *Disc.* in Isle of Anglesea. 1845. Peter Spence *pat.* method of manufacturing alum by treating burnt shale and iron pyrites with sulphuric acid.

Aluminium 1754. Andreas Sigismund Margraff (1709–82) (Ger) proved aluminium to be a distinct earth from lime. 1825. Hans Christian Oersted (1777–1851) (Dan) first *isol.* aluminium as chloride. 1827. *Disc.* F. Wöhler (Ger). 1833. Michael Faraday (1791–1867) first *prod.* aluminium electrolytically. 1856. *Prod.* simplified by H. St. Claire-Deville (Fr). 1882. Aluminium electrolysed by Charles Martin Hall (1863–1914) and Paul Louis Toussaint Hérault (Fr) (1863–1914), who produced it in an electric furnace from cryolite, in 1886. Sodium process of production *inv.* Hamilton Y. Castner (U.S.).

Amalgam 77 B.C. Gold amalgam *desc.* by Pliny. 27 B.C. Gold amalgam *desc.* by Marcus Vitruvius Pollio (*c.* 50–26 B.C.).

Ambulance *c.* 1500. Litters used for carrying wounded soldiers. 1790. Saint Sauveur (Fr), *inv.* Hammock carriage. 1800. Jean Dominique Larrey

(Fr), of Beaudéan, Pyrenees, and Pierre Francois Percy (Fr), *inv.* "Flying Ambulance" with springs.

Americium (element) *Disc.* 1952–53 by Glen T. Seaborg (U.S.) and A. G. Liorsa (U.S.). *See also* Elements Berkelium, Californium, Einsteinium, Fermium and Mendelevium.

Ammonia (gas) 1315. First noticed scientifically by Majorcan philosopher Raymond Lully (Lull) (1232–1315), who was stoned to death by townsfolk of Parma for his discovery. 1677. Johann Kunckel (1630–1703) (Ger) *desc.* aqueous solution of Ammonia. 1774. Joseph Priestley (1733–1804) *isol.* ammonia gas. 1787. M. van Marum (Hol) and A. Paeto van Troostwijk (Hol) succeeded in liquefying ammonia. 1855. Ammonia gas utilized as a refrigerant by Carré (Fr). Ammonia refrigeration process improved by Reece (1869) and Linde (Fr) (1880). 1909. Ammonia first synthesized by Fritz Haber (Ger) (1868–1934).

Ammoniaphone *Inv.* 1883. Dr. Carter Moffatt.

AMONTONS, Guillaume (1663–1703) (Fr) 1700. *Imp.* Galileo's thermoscope. *See* Thermometer.

AMPÈRE, André Marie (1775–1836) (Fr) 1820. *Disc.* magnetic effect of electricity passing through a coil of wire. *See* Electro-magnet. *See also* J. S. C. Schweigger, Oersted, Nobili, Pouillet and William Thompson.

Ampmeter (Ampéremeter) 1820. Principle of ampmeter

inv. Hans Christian Oersted (1777–1851) (Dan). 1884, ampmeter introduced in electrical engineering.

Amputation (surgery) 1679. Flap method *inv.* Dr. Lowdham, of Exeter.

Amylene (chemistry) 1844. First procured from potato spirit by Antoine Jérôme Balard (1802–76) (Fr), of Paris.

Anæmia 1926. Use of liver in treatment of *disc.* by Minot and Whipple.

Anagram *Inv.* by Lycrophon, 280 B.C.

Analysis 1830. J. B. A. Dulong (Fr) perfected incineration method and absorbing gases *prod.* to determine percentage of carbon, hydrogen, etc., in organic compounds. 1849–53. Sir Edward Frankland devised new analytical and synthetical principles.

Anatomy, Pathological Created by Coiter (Fl. 1534).

Anatomy, Comparative 1675. Term *intro.* by Nehemiah Grew (1641–1712).

Anchor 592 B.C. Supposedly *inv.* Anacharis the Scythian. 569 B.C. Fluke, or second tooth added. (A.D.) 578. Anchors first forged in England. *c.* 1840. Porter *inv.* the feathering anchor. *c.* 1850. John Trotman *imp.* Porter's anchor.

Anemometer 1667. Anemometer *inv.* Dr. Crone. Pendulum type *inv.* Robert Hooke (1635–1703). 1753. Pendulum type *inv.* Marquis Poleni (Fr). 1744. Pendulum type re-invented by Roger Pickering of Dartford, Kent. 1775. Fluid-type anemometer *inv.* Dr. Lind. U-tube type *inv.* Pierre Daniel

Huet (Fr). 1709. Another type *inv.* Wolfius. 1843–6. Dr. Robinson, of Armagh *inv.* rotating A. 1844. Pressure-plate anemometer *inv.* by Osler and Dr. Whewell and fitted to weathercock of Royal Exchange, London, 1862.

Aneroid *See* Barometer.

ÅNGSTRÖM, Anders Jonas (1814–74) (Swed) 1868. Evolved the Ångström unit. *See* Spectrum, also Ångström Unit.

Ångström Unit 1868. Evolved by Anders Jonas Ångström (1814–74) (Swed) to measure Fraunhöfer's spectrum lines *q.v.*

Aniline 1826. *Disc.* Unverdorben (Ger) among products *obt.* from natural indigo, *q.v.* 1850. *Obt.* from benzol by (Sir) William Henry Perkin (1838–1907) while attempting to synthesize quinine, *q.v.* Aniline first prepared by Fritsche (Ger), 1841.

Animalculæ *Disc.* Antoine van Leeuwenhock (Leeuwenhoek) (1632–1723). (*Pub.* in his *Arcana Naturae*, 1696.)

Anthracine 1832. *Disc.* in coal-tar by Dumas and Laurent. 1868. *Obt.* from Alizarine by Graebe and Liebermann.

Anthrax bacillus 1849. *Disc.* Aloys Antoine Pollander. 1877. Bacillus life-cycle *dem.* by Robert Koch (Ger).

Anthropology Branch of science founded *c.* 1805 by Johann Friedrich Blumenthal (1752–1840) (Ger).

Anticyclone Atmospheric high-pressure area. Name coined by Sir Francis Galton and introduced by him in his "Meteorographica," 1861.

Antifibrin (drug) 1886. Synthesized.

Antimony (metallic element) 1640. First *isol.* and *desc.* German monk Basil Valentine (Johann Thölde).

Antinomies (of aggregates) theory 1897. First *prop.* Burali-Fort (It).

Antipodes, Idea of *c.* 388 B.C. First conceived by Plato (429–347 B.C.).

Antipyrene (drug) 1883. Synthesized.

Antiseptic 1865. First used by Joseph, Baron Lister (1827–1912).

Antitoxin 1890. Word *intro.* by Emil von Behring (1854–1917).

Apollonicon (musical instrument) 1817. *Inv.* Flight and Robson, of Westminster, London.

Appendicitis (surgical) First established as a definite lesion and named by Reginal Heber Fitz (1843–1913) (U.S.), of Boston, Mass.

APPERT, Francois (1750–1841) (Fr) 1795. *Appl.* sterilization process to food by bottling and canning; heating and sealing.

APPLEGARTH, Augustus (1788–1871) 1818. *Pat.* cylinder printing machine.

APPOLT, John George (1800–54) 1825. *Inv.* chamber gas-producing retort.

Aquatint *Inv.* by German artist Le Prince (*b.* 1723).

ARAGO, Dominique Francois (1786–1853) (Fr) *Dis.* chromatic polarization of light. *See* Optics.

Arc, Singing 1906. Phenomenon first *obs.* by William Duddell, F.R.S. (1872–).

Arc, Electric 1802. *Disc.* by Sir Humphry Davy.

Arc, Undamped 1903. *Inv.* by Valdemar Poulsen (Dan) for transmitting speech by radio and *dev.* by H. P. Dwyer (U.S.) to transmit on 500 metres.

Arch 1550. Strength of arch measured by Andrea Palladio (It) (1518–80). Palladio's measurements reduced to correct formulae by Derand and La Hire (Fr), 1695.

Archery Ascribed to Apollo (*leg.*), who communicated weapon to Cretans. *Intro.* England, pre-440. First notice, Genesis xxi, 20.

ARCHIMEDES of Syracuse (*c.* 287–212 B.C.) Made first attempt to measure specific gravity, *q.v. Inv.* screw. (*See also* Burning-mirror.)

ARGAND, Aimé (1755–1803) (Swit) *b.* Geneva. 1808. *Inv.* Gas-burning lamp bearing his name.

Argand Lamp 1789. Iron-chimney Argand lamp *inv.* Aimé Argand (1755–1803) (Swit). Glass chimney Argand lamp *inv.* Lange. 1808. Argand burner used by Samuel Clegg as gas-burner. *See also* Lamps, Gas-burners.

Argon (element) 1894. *Disc.* Prof. William Ramsay.

ARISTOTLE 384–322 B.C. Philosopher and logician.

Arithmetic 600 B.C. said to have been introduced into Greece from Egypt by Thales the Miletian (640–550 B.C.). Oldest treaties on arithmetic by Euclid (Gr) (Fl. *c.* 400 B.C.) in his 7th, 8th and 9th books of Elements. 130. Sexigesimal system of Ptolemy in use. 6th cent.

Decimal notation (9 digits and zero) used in India. 900. *Intro.* Arabia. 980. *Intro.* Europe. 991. *Intro.* France by Gebert. 1050. *Intro.* Spain. 1253. *Intro.* England. 1863. Tonal system with 16 as basis *Pub.* by Nyström. *See also* Decimal system.

ARKWRIGHT, Sir Richard (1732–92) 1769. Accredited with *inv.* the roller-spinning water-loom. (Machine actually *inv.* Thomas Highs, and process by Lewis Paul.)

Armillary Spheres *c.* 255 B.C. *Inv.* Eratosthenes (276–196 B.C.) (Gr). 1562. Automatic armillary sphere *desc.* and *illus.* by J. Taisner (Ger).

ARMSTRONG, Sir William G. (later Lord Armstrong) (1810–1900) *Inv.* Hydraulic accumulator, "Armstrong-Gun," etc.

Arquebus Mentioned as early as 1476. 1485. Yeomanry armed with the arquebus. *See also* Arms.

ARRHENIUS, Svante August (1859–1927) (Swed) 1886. *Prop.* theory of Electrolytic Dissociation.

Arsenic (element) *c.* 800, Geber (Giaber, or Yeber) (Arab) (9th cent.) *Disc.* Arsenic oxide and obtained it from burning the mineral realgar (arsenic sulphide). 1733. Kennig Brandt (1674–1768) (Ger) first examined metallic arsenic with precision. 1775. Carl Wilhelm Scheele (1742–86) (Swed) *disc.* Arsenic acid.

ARSONVAL, Jacques Arsène d' (1857–1940) (Fr) *Inv.* Mirror Galvanometer.

Artesian Well *See* Well.

Asbestos *c.* 170. Mentioned by

Greek religious antiquary Pausanius, as "Carpasian Linen." *c.* 1500. Asbestos spun at Venice, 1684. First mentioned in modern times at meeting of Royal Society, London. Also mentioned by Herodotos and Pliny.

Ascorbic Acid 1928. Isolated by Györgi Szent from oranges, lemons and cabbages. King and Waugh later obtained it in crystal form.

ASELLI, Gasparo (It) (1581–1626) 1627. *Disc.* the lacteal ducts.

Asphalt 1595. Sir Walter Rayleigh visited asphalt lake, Trinidad. 1712. asphalt *disc.* Greek Doctor at Neuchâtel, Switzerland. 1800. After bitumen, used for pavements in France. 1849. First road laid with Asphalt at Val de Cranersby by Swiss engineer Merian.

Aspirator (surgery) *Inv.* Armand Trosseau, Paris (1801–67). Another type by Pierre Carl Édouard Potain (1825–1901) (Fr). *Imp.* by Georges Dieulafoy (1839–1911) (Fr).

Astatine (element) 1940. *Disc.* K. R. Mackenzie, E. Segré and D. R. Corson, in U.S.

Astigmatism 1827. Recognized by Sir George Airy (1831–81), who corrected the fault in his own eyes.

ASTON, F. W. (1877–1941) 1931. Announced detection of U238 in the spectrograph of uranium hexafluoride.

Astrolabe *Inv.* and used by Hipparchus (*c.* 160–125 B.C.). Used by Ptolemy, Claudius (2nd cent. B.C.). 1513. *Imp.* by Fabricius.

Atmolysis (gas separation) 1863. Method *disc.* Prof. T. Graham, F.R.S.

Atmosphere, Composition of 1777. Proved to be composed of oxygen and Nitrogen by Karl Wilhelm Scheele (Swed) (1735–86). 1801. Proportions of constituent gases (oxygen and nitrogen only) found by Joseph-Louis Gay-Lussac (Fr) (1778–1850) and Baron Alexander von Humboldt (Ger) (1769–1859).

Atolls 1837. Theory of atolls and coral reefs *prop.* by Charles Darwin (1809–82). 1950. Darwin's theory confirmed by 4,222 ft. borehole on Eniwetok Atoll, Pacific Ocean.

Atom 1913. Niels Bohr (Dan) (*b.* 1885) solved problem of structure of hydrogen atom and its spectrum, thus reconciling Rutherford's nucleus theory with the quantum theory of energy. 1919. Sir Ernest Rutherford (later Lord Rutherford) (1871–1937) (N.Z.) first artificially disintegrated an element and transmuted many lighter elements.

Atomic theory 1777. Karl Friedrich Wenzel (Ger) (1740–93) undertook early work on theory. 1808. Theory *prop.* by John Dalton (1776–1844). (*See also* his Law of multiple proportions.) 1811. Amadeo Avogadro *prop.* Hypothesis on atoms and molecules.

Atomic Weights 1858. Consistent system of Atomic Weights pioneered by S. Cannizarro (It).

Atomic Heating Plant ("B.E.P.O.") 1951. First plant opened at Harwell, England.

Atomic Nuclei 1932. First fissioned by John Cockroft (1897–).

Attraction (physics) 1520. First *desc.* by Nicholaus Copernicus (Ger) (1473–1543). Theory expanded by Johann Kepler (Ger) (1571–1630). 1687. Sir Isaac Newton (1642–1727) *prop.* his theory of attraction.

Audion *See* Valve, Thermionic.

Auger (tool) 600 B.C. In use in Ancient Egypt.

Aureomycin (drug) 1947. *Disc.*

Auroræ 1735. Relation to earth's magnetic field *disc.* John Hadley (*d.* 1744). *Desc.* Aristotle ("De Meteoris," lib. 1, c. 4, 5).

Autocycle Wheels *See* Wheels, Autocycle.

Autogiro 1907. Man first lifted by rotating-wing aircraft. *Inv.* Juan de la Cierva (Sp), 1922. 1923, Jan. 9 First flight (200 yards). 1923, Jan. 21 Flight of 2½ miles in 3½ minutes. (*See also* Aircraft.)

Automata 400 B.C. Flying dove made by Archytas of Tarentum. 1264. Friar Bacon said to have made bronze head that spoke. 1649. Jean Pierre Camus (Fr) (1582–1652) made model coach and horses with footmen, pages and passenger for Louis XIV—when a child. 1738. Jacques de Vaucanson (Fr) (1709–82) made artificial duck that ate, drank and quacked; also an automatic flute-player.

Automatic Transmission *See* Hydraulic coupling and Torque converter.

Autotypography (metal plate from drawings) 1863. Process made known by Wallis.

AVELING, Thomas (pioneer traction-engine builder) 1858. Built his first engine. *See* Traction-engine.

AVOGADRO, Count Amadeo (It) (1776–1856) 1811. *Prop.* hypothesis on atoms and molecules.

Axe (tool) 1240 B.C. *Inv.* (*Leg.*) by Daedalus of Athens.

Axle, Stub 1828. *Inv.* by Gough.

B

BABBAGE, Charles (1791–1871) 1812. *Des.* mechanical computing "engine."

BABBITT, Isaac (1799–1862) 1839. *Inv.* anti-friction alloy now bearing his name.

BABCOCK, G. H. (1832–93) 1867. With Wilcox *inv.* tubular steam boiler.

Backgammon 1224 B.C. *Inv.* Palamedes (Gr. *Myth.*). Also said to have been *inv.* in Wales before its conquest.

Bacteria 1864. First *disc.* Louis

Pasteur (Fr) (1822–95). (*See* also Virus.)

BÆKELAND, Leo Hendrik (Bel) (1863–1944) 1909. *Inv.* earliest plastic—Bakelite.

Baffin's Bay 1616. *Disc.* William Baffin (*c.* 1584–1622).

Bagpipes A.D. 51. Greek sculpture in Rome depicts bagpipe player dressed as highlander of today. Nero said to have played on bagpipes.

BAINBRIDGE, William 1803. *Inv.* the flageolet.

Baize 1660. First manufactured in England at Colchester.

Bakelite (plastic) 1909. *Inv.* Leo Henrik Bækeland (Bel) (1863–1944).

Balance 5000–4000 B.C. Scales with $3\frac{1}{2}$ in. beam of red limestone and set of weights for weighing gold-dust found in tomb at Naqada, Egypt. 3000 B.C. Balances widely used in trade. 1350 B.C. Balances to weigh a shekel (7–14 grammes to an accuracy of one per cent in use. A.D. 80. Steelyard balance in use in Rome. 1694. First *illus.* of spring balance. 1772 John Sebastian Clais (English *pat.* No. 1014). *Inv.* "Index Balance" with single counterpoise and index-dial instead of weights. *c.* 1780 Torsion balance *inv.* Augustin Coulomb (Fr) (1736–1806). 1658. Watch balance-spring *appl.* by Robert Hooke.

BALARD, Antoine Jérôme (Fr) (1802–76) 1826. *Disc.* element bromine.

Ballistae (arms) 100 B.C. Mentioned 2 Chronicles xxvi, 15.

Ballistocardiagraph (medical) *Inv.* by Henderson.

Ball-joint (pipe) 1880. *Inv.* E. P. Monroe (U.S.).

Balloon 1670. First idea for balloon *pub.* by Jesuit Francesco de Lana. 1709. Jesuit Father Laurenço de Gusmão (1686–1724) (Port) *b.* Santos, Brazil; *d.* Toledo, Spain. *Inv.* hot-air balloon. 1783, June 5. Montgolfier brothers (*q.v.*) sailed unmanned hot-air balloon at Annonay, nr. Lyons, France. Aug. 27. J. A. Charles (Fr) made first hydrogen balloon, which ascended from the Champ de Mars, Paris unmanned. Oct. 21. Pilatre de Rozier (Fr) and Marquise d'Arlandes (Fr) first men to ascend in free flight. Attained height of 3,000 ft. and travelled from La Muette, Paris to Gonesse, 45 miles. Dec. 1. Charles and Roberts (Fr) ascended from Tuileries Palace grounds to a height of 2 miles. Dec. 28. First man to ascend in New World: James Wilcox (U.S.). 1784, Jan 19. 5,000,000 cu. ft. hot-air balloon rose to 3,000 ft. with 7 passengers, including Joseph Montgolfier and Pilatre de Rozier. Mar. Jean Pierre Blanchard (Fr) rose to 9,000 ft. at Paris. April. Mm. Morveau and Bertrand rose 13,000 ft., and travelled 18 miles, in 25 minutes at Dijon, France. July. First ballooning accident (not fatal), over Paris. Sept. First balloon voyage over Britain by Vincent Lunardi (Neapolitan) from London to Standon, Hertfordshire. 1785, Jan. 7. Blanchard (Fr) and Dr. Jeffries (U.S.) crossed English Channel from Dover to Calais. June 15.

Rozier and Romain (Fr) try to cross Channel from Boulogne. Rozier killed and Romain survived 10 minutes (first flying fatalities). 1803. First night flight: Count Zambeccari, Dr. Grassati and Pascal Andreoli (It) from Bologna to Istrian Coast. 1836. Messrs. Green, Holland and Monck Mason fly 500 miles in 18 hours (London–Weilburg, Germany). 1862. James Glaisher reached height of 7 miles. 1906, Feb. 20. Mrs. Griffith-Brewer the first woman to cross English Channel by air—in a balloon. (*See also* Aircraft)

BALLOT, Christopher Hendryk Buys (Hol) (1817–90) *Prop.* Law bearing his name (*q.v.*).

Balmer's Series (spectroscopy) 1885. Formula discovered for lines of wave-length series by Swiss schoolmaster Johann Jacob Balmer. *See also* Rydberg (Swed).

Bank, Savings 1798. First *prop.* Rev. Josiah Smith, of Wendover, Buckinghamshire. 1810. Penny savings bank *est.* at Paisley (Penny Bank Friendly Society) by Rev. Henry Duncan. 1817. 70 savings banks in England and Scotland; 4 in Wales and 4 in Ireland.

Barbed Wire 1875. First *inv.* and used for cattle fencing.

BARBER, John 1791. *Pat.* power unit working on gas-turbine principle. (*See* Gas turbine.)

Barge, "Dracone" 1956. *Inv.* Prof. W. R. Hawthorne, of Cambridge.

Barium (element) 1803. *Disc.*

1803, Karl Wilhelm Scheele (Swed) (1742–86).

BARKER, Dr. Robert *c.* 1745. *Inv.* water turbine bearing his name.

Barker's Mill *c.* 1745. *Des.* Dr. Robert Barker. Desaguliers (Fr) later made a model of this turbine.

BARNETT, William (U.S.) 1838. *Pat.* two-pump gas-turbine.

Barometer 1643. *Inv.* Evangelista Torricelli (It) (1606–47). 1648. Barometer first used as altimeter by Blaise Pascal during experiments on Puy de Dôme. 1844. Aneroid barometer *inv.* Lucien Vidie (Fr) (1805–66), but also ascribed to A. A. S. Conté (Fr) 1798. 1668. Wheel barometer *inv.* Dr. Robert Hooke (1635–1703). 1657. Water barometer *inv.* Otto Guericke, of Magdeburg. It was 33 ft. long and was re-*inv.* by Prof. Daniell in 1830. 1695. Pendant barometer *inv.* 1657 marine pendant barometer *inv.* 1805. Air pressure phenomenon *disc.* by Pascal placed on qualitative basis by Pierre Simon Laplace (Fr). 1882. Glycerine barometer *inv.* by geologist Jordan, of London.

Barrel-organ *See* Pianoforte.

BARSANTI, E. (It) 1857. *Inv.* gas engine (*q.v.*).

BARTON, John (n.d.) *Inv.* Metallic Piston packing. (*q.v.*)

Bassoon (musical instrument) *Inv.* Alfranio of Ferrara 1530.

Bathing Machine *Inv. c.* 1790. First used at Margate, Kent.

Baton, Conductor's 1820. *Intro.* England by Ludwig Spohr (Ger) (1784–1859).

Battering-ram 441 B.C. Used

by Pericles (A.G.). 330 B.C. Diades, Greek engineer *inv*. roller-mounted battering-ram. **Battery, Electric** 1800. "Voltaic Pile" *inv*. Alessandro Volta (It) (1745–1827). 1802. "Dry Pile" *inv*. Georg Bernhard Behrens (1775–1813); re-*disc*. Guiseppe Zamboni (It) (1776–1846). Wet battery *inv*. John Frederic Daniell (1790–1845) (1 volt per cell). 1839. Battery *inv*. Sir Robert William Grove (1811–96) (1.78 volts per cell). 1841. Wet battery *inv*. Robert Wilhelm Bunsen (Ger) (1811–99). 1859. Potassium bichromate battery *inv*. Grenet (Fr), also by Poggendorf (Ger). 1867. Wet battery *inv*. Georges Leclanché (Fr) (1839–82). 1873. Zinc-mercury battery *inv*. Latimer Clark (the "standard cell"). 1880. Dr. De la Rue *inv*. silver chloride and zinc battery. (A thermo-electric battery *inv*. by A. C. Becquerel.) 1860. Benoist, of Paris, battery *pat*. in England by Mennons.

Battery, Floating (Naval armament) 1780. *Inv*. Claude Éleonore le Michaud, Chevalier d'Arçon (Fr) (1733–1800). Ten were made and used to attack Gibraltar in 1782.

BATTIN, Joseph (U.S.) 1856. *Pat*. vertical boilered, oscillating cylindered steam road carriage.

BAUER, Georg See Agricola, Georgius.

BAYER, Adolf von (Ger) 1871 to 1887. *Disc*. phtalein dyes.

Bayonet (arms) 1641. Originated in Bayonne. (Bayonet had to be removed from musket before firing.) 1691. Socket bayonet *inv*. Br. Gen. Mackay 1703. Bayonet adopted for all French infantry. 1707. Socket bayonet *inv*. De la Chaumette (Fr).

Beams, Wrought-iron 5th cent. B.C. Used as cantilevers to support heavy strucures: *e.g.* Pantheon, Athens. Beams 5 in. × 12 in. × 15 ft. used to cap stone pillars of temple at Agrigento. 1846. *Intro*. as result of carpenters' strike in France.

Beams, Tensile Testing of 1815. Dulin (Fr) tested wooden beams. 1820. Duleau (Fr) tested built-up iron beams. 1826. Navier (Fr) tests breaking load of beams. 1833. Lamé and Clapeyron (Fr) formulated equations to determine stresses in continuous beams. 1847. Eaton Hodgkinson experimented with cast-iron beams and chose I-beam as the strongest. Clapeyron published his 3-moment equation. (*See also* Tensile testing machines.)

Bearing, Antifriction 1839. *Inv*. I. Babbit (1799–1862). 1710. De Mondran (Fr) *des*. carriage with anti-friction bearings.

Bearing, Ball A.D. 12–41. Ball bearing thrust race found on Roman galley retrieved from Lake Nemi in 1928. (Bronze balls between wooden discs.) 1772. Ball bearing *desc*. by C. Varlo. 1780. Ball bearing thrust race with 2½ in. cast-iron balls fitted to post windmill at Sprouston, Norfolk. (Race now in Bridewell Museum, Norwich.) 1794. First *pat*. for ball bearing Phillip Vaughan to carry radial load. 1849. Edward Bancroft (U.S.) *inv*. self-aligning ball bearing. 1862. A. L. Thirion *intro.* ball bearing to bicycle. 1868. Ball bearing (first) fitted to bicycles by Cowper. 1878. Ball bearing fitted to bicycles by Daniel Rudge. 1901. Ball-thrust bearing *inv.* E. G. Hoffman.

Bearing, Dry 1954. Plastic (Polytetrafluorethylene or P.T.F.E.) bearings announced as having a dry friction co-efficient less than that of a normally lubricated bearing. (280° C. temperature permissible.)

Bearing, Roller 484–424 B.C. Herodotus mentions use of rollers for land transport of ships, etc. 330 B.C. Engineer Diades (A.G.) *inv.* (?) roller-mounted battering-ram: 1st cent. B.C. Wooden roller bearings found on hubs of Dejbjerg wagon wheels. *c.* 1470. Leonardo da Vinci (1452–1519) mentions rollers as friction reducers. 1550. Georg Agricola (Ger) (1494–1555) and Augustin Ramelli (It) (1531–90) refer to and *illus.* machines with roller bearings in their books, *c.* 1715. Henry Sully made ship's chronometer with anti-friction roller bearings. (Usually referred to as *inv.* of roller bearings.) 1734. British *Pat.* No. 543: Jacob Rowe — friction wheels with axles bearing on other wheels. 1750. Garnett *inv.* roller bearing. 1787. L. Garnett, of Gloucester took out first *pat.* for a roller bearing. 1787. George Watkin (*Pat.* No. 1602) *inv.* roller bearings with roller cage. Also included use of cylinders and cones as rollers. 1800. Thrust roller bearing *inv.*

with coned rollers. 1898. Tapered roller bearing *inv.* H. Timken and R. Heinzelmann.

Bearing, Self-aligning 1842. J. G. Bodmer (Swit) (1786–1854) *inv.* self-aligning bearing. 1849. Edward Bancroft (U.S.) *inv.* ball self-aligning bearing.

Bearing, Self-lubricating 1750. Used by J. H. Harrison in his maritime chronometers.

BEAU DE ROCHAS, Alphonse (Fr) (1815–91) 1862. *Prop.* four-stroke compression cycle for internal combustion engines.

BECHER, Johann Joachim (Ger) (1635–82) 1669. *Prop.* "terra pinguis" (oily earth) theory for the combustability of chemical compounds.

BEEKMANN, Isaac (Hol) (1588–1637). Deduced dynamically the law of falling bodies 13 years before Galileo. (His journal lost until beginning of 20th cent.)

BECQUEREL, Antoine César (Fr) (1788–1878) *Inv.* thermoelectric battery; and in 1850 an electric pyrometer.

BECQUEREL, Antoine Henri (Fr) (1852–1908) 1898. *Disc.* radio-activity.

Bed *c.* 1587–1375 B.C. (18th Dynasty) Bed with head end higher than foot end used in Ancient Egypt. 1851. One of the earliest metal beds exhibited at the Great Exhibition 1851.

Beef, Extract of 1847. Process *inv.* Baron Justus Frieherr von Liebig (Ger) (1803–73).

Bees 1670. Sex of *disc.* Swammerdam (Hol). 1760. Bees *intro.* America.

"Beetleware" (urea formaldehyde) *See* Plastics.

BELISARIUS, General (A.R.) (*c.* 505–65) *Inv.* floating mill.

BELL, Alexander Graham (1847–1922) 1876. *Inv.* speaking telephone. (Elisha Grey (U.S.) applied for *pat.* the same day.)

BELL, Henry (1767–1830) 1812. Built steamship *Comet*, which plied the River Clyde from Glasgow to Greenock.

BELLIS, George Edward (1838–1909) 1889. *Inv.* system of forced lubrication for high-speed steam engines.

BELON, Pierre (Fr) (1517–64) Naturalist who first recognized cetaceans as mammals.

Bellows *c.* 2300 B.C. Suggested at Tuyer smelting furnaces found at Ur of the Chaldees. *c.* 2000 B.C. Hittite forges used bellows. 4th cent. B.C. Double-acting piston bellows for continuous blast used in China. *c.* 569 B.C. Bellows *inv.* (*Leg.*) Anacharis of Scythia. A.D. 4th cent. House-bellows *desc.* by Decimus Magnus Ausonius (A.R.) (310–95). 12th cent. Forge bellows with wooden top and bottom boards and leather flap valves *inv.* 1669. Brothers Martin and Nicholas Schelhorn, of Coburg, *inv.* wooden bellows. 1550. Made at Nuremburg for smelting and organ-blowing. 1621. *Inv.* ascribed by Beckmann to Lewis Pfannenschmidt, of Ostfeld, near Goslar, Germany.

BENZ, Carl (Ger) (1844–1929) 1884. *Inv.* a gas-engine and an internal combustion engine.

Benzene 1825. *Disc.* Michael Faraday.

Benzene hexachloride 1825. First prepared by Michael Faraday.

Benzole 1825. *Disc.* by Michael Faraday in oils. 1849. *Disc.* by C. B. Mansfield in coal-tar.

B.E.P.O. *See* Heating plant, Atomic.

BERGMANN, Torbern (1735–84) 1778. With C. W. Scheele *disc.* element molybdenum.

Beri-beri 1885. Japanese medical officer Takaki *disc.* disease to be of dietary origin.

Berkelium (element) 1949. *Disc.* by Glen T. Seaborg (U.S.) in conjunction with A. G. Liorso (U.S.).

BERLINER, Emil (1851–1929) 1887. *Inv.* gramophone (disc) sound-reproducing machine.

BERNARD, Claude (Fr) (1813–78) 1848. *Disc.* glycogen (animal starch). Also suggested use of hypodermic syringe.

BERNOULLI, Daniel (Fr) (1700–82) (son of Jean Bernoulli) 1745. *Prop.* the principle of areas or principle of the moment of momentum.

BERNOULLI, Jacques (Fr) (1654–1705) (brother of Jean) Solved problem of isoperimeters. (*See* Calculus of variations.)

BERNOULLI, Jean (Fr) (1667–1748) (brother of Jacques) 1715. Elucidated the principle of virtual velocities or work. *Disc.* exponential calculus.

BERTHELOT, Pierre Eugene Marcelin (Fr) (1827–1907) *c.* 1870. *Prod.* acetylene from carbon and hydrogen at temperature of electric arc.

BERTHOLLET, Claude-Louis (Fr) (1748–1882) *Disc.* hypochlorites and chlorites.

Beryllium (Glucinum) (element) 1798. *Disc.* Vauquelin. 1828. First *obt.* as a metal by Friederich Wöhler (Ger) (1800–82).

BERZELIUS, Jöns Jakob (Swed) (1799–1848) *Disc.* elements selenium, thorium, silicon and zirconium.

BESSEMER, Sir Henry (1813–98) 1860. *Inv.* steelmaking process later *imp.* in the 1870s by Sidney Gilchrist Thomas.

BESSON, Jacques (Fr) (*c.* 1550) Author of *Theatre of Instruments, pub.* 1582.

BICKFORD, William of Tuckingmill, near Camborne, Cornwall. 1831. *Inv.* miners' safety fuse.

Billiards 1571. *Inv.* artist Henrique Devigne (Fr). 1578. Tables kept by Lombards in Holland. 1827. Slate bed *intro.* into England.

Binominal Root 1550. Term first used by Robert Ricorde.

Binominal Theorem 1676. First mentioned as his *inv.* by Sir Isaac Newton in a letter to Leibnitz. 1770. Theorem proved by Leonhard Euler (Swit) (1707–83).

Binominal Nomenclature (natural history) Ascribed to Linnaeus. Idea proposed by Bauhin (Swit). Joachim Jung, of Lubecl (1517–1657) named plants with two names—noun and adjective.

Birds 875. Anatomy of birds

first *desc.* by Thebit Ben Corrah (Arab).

BIRKELAND, Kristian (1867 –1917) 1903. With Samuel Eyde (1866–1940) *inv.* nitrogen fixation process.

Bismuth (element) 1546. First *desc.* by Georg Bauer (Georgius Agricola) (Ger) (1494–1555), also by Johann Thölde (Basil Valentine).

BLACK, Joseph (1728–99) 1755. Showed difference between caustic and mild alkalis, and many other *discs.* in physics.

BLACKBURN, A. B. 1877. *Des.* first liquid fuel steam road vehicle in great Britain—a dog-cart.

Blagden's Law (relating to freezing of liquids) 1788. First *obs.* by Charles Blagden. 1861. F. Rüdorff announced law as new *disc.* 1871. De Coppet (Fr) re-*disc.* Blagden's work. 1882. François Marie Raoult (Fr) (1830–1901) extended researches on Blagden's law.

Blankets 1337. First made by Blanket Bros., of Bristol.

Blast Furnace 1621. First coal-fired iron-smelting furnace claimed by Dud Dudley (1599–1684). 1753. First coke-fired furnace erected at Coalbrookdale by Abraham Darby Jnr. 1758. First blast furnace erected by John Wilkinson (1728–1808) at Bradley, Bilston. 1821. Pre-heating of air-blast *inv.* Neilson of Glasgow. 1845. J. P. Budd *pat.* hot blast from furnace waste gases. 1860. Tower-type blast furnace *inv.* Edward Cowper. (*See also* Blowers, air and gas.)

Blast, Hot (smelting) 1828. First tried at Clyde Ironworks, by Neilson. 1860. Cowper *intro.* hot blast at Middlesbrough.

Bleaching 1500. Blueing for bleaching *intro.* G.B. from Holland. 1756. Dilute sulphuric acid used for bleaching by Hulme. 1785. Claude-Louis Berthollet (Fr) (1748–1822) *intro.* chemical bleaching with "Eau de Javal." 1789. Berthollet *inv.* chlorine and alkali bleaching liquid. 1799. Charles Tennant, of Glasgow (1768–1838) heard from James Watt of the Berthollet process and *imp.* it by *inv.* bleaching powder.

Blind, Printing Systems for the 1784. First system *inv.* Haüy, of Paris. 1832. Sir Charles Lowther printed by Haüy's method. 1834. Gall's system *inv.* 1843. Braille (Fr) *inv.* system. 1876. Dr. Thursfield, of Shrewsbury *inv.* typewriter for use of blind. (*See* Typewriters.) Other systems: Lucas's, Frére's and Moon's.

Blocks, Printing A.D. 868. Earliest Chinese books appear with blocks.

Blood 1854. Welcker *disc.* method of finding total amount in body. 1889. Constitution of at high altitudes *disc.* by Viault in the Peruvian Andes. 1911. Existence of blood groups proved by Landsteiner, more groups being later found by Dungern and Herzfeld. 1940. Rhesus factor *disc.* by Weiner.

Blood, Circulation of 1553. M. Servitus (Sp) published clear account of pulmonary circulation of blood. Reputed as *disc.* by Honore Fabri (1607–88). 1615. *Disc.* William Harvey (1578–1657) but not announced until 1628.

Blood, Pressure of Determined by Spallanzani (It) (1729–99). (*See also* Sphygmomanometer.) 1905. Korotov (Rus) *inv.* ausculatory method of checking.

Blood-pump, Mercurial *Inv.* by Karl Ludvig (Ger) (1816–95) and perfected by E. F. W. Pflüger (Ger) (1829–1910).

Blowers (Air and Gas) 2nd cent. B.C. Water-powered blowers in use in China. *c.* 1500. Italian "trompe" superceded bellows (*q.v.*) for iron production. (Blast produced by suction of water running down a chute.) 1866. Rotary blowers *inv.* J. D. Rootes (U.S.) and produced at Connersville, U.S. by P. H. and F. M. Rootes. *See also* Supercharger. *c.* 1760. John Smeaton made the first blower for Carron Ironworks.

Blowpipe 1750. *Intro.* by Axel Frederick Cronstedt (Swed) (1722–65) for isolation of nickel from its ores.

BLUMENTHAL, Cellier 1808. *Inv.* and constituted first rectifying column for separating alcohol from wine. *See* Distillation.

Boat-lowering Gear 1856. *Inv.* by Charles Clifford. *See also* Ship.

BODE, Johann Elert (Ger) (1747–1826) *Prop.* astronomical law named after him.

Bode's Law (astronomy) 1778. Enunciated by Johann Bode (Ger) (1747–1826), but law usually ascribed to Prof. Johann Daniell Titius (Tietz) (1729–96), of Wittenberg or to Christian Wolff (1679–1754).

BODMER, John G. (Swit)

(1786–1854) Prolific mechanical *inv. See* Gear-cutting, Tyre rolling-mill, Self-aligning bearing and Automatic stoker.

BOEHM, Theo (Ger) 1823. *Inv.* " new " system of keying flutes.

BOETHIUS, Anicius Torquatus Severinus (A.R.) (470—526) Writer on mathematics, astronomy, and music.

Bogie (railway engineering) 1812. *Pat.* by William Chapman in England.

Boiler, Steam pre-1725. Copper and wrought-iron plate haystack boiler *inv.* 1780. Multi-tube boiler *des.* by Charles Dallery (Fr) for steam road vehicle. 1786. "Cornish-type" boiler said to have been *inv.* by Oliver Evans (U.S.). 1791. Vertical boiler *pat.* Nathan Reed, of Salem, Mass., U.S. 1808 (1812?). Cornish boiler *inv.* Richard Trevithick. 1823. Jacob Perkin *inv.* flash-steam boiler working at 1,500 p.s.i. 1825. Eve *inv.* vertical, tubular boiler. 1826. Multitubular type *inv.* by Sir Goldsworthy Gurney (1793–1875) and used in his steam road coach. 1827. Marc Seguin (Fr) *pat.* multi-tube boiler and applied it to locomotive, 1829. 1828. Booth *inv.* multi-fire-tube "loco-type" boiler with Stephenson. 1828. Steendrup *inv.* vertical boiler. 1830. Nott *inv.* vertical boiler. 1833. Col. Francis Macerone *des.* water-tube boiler for road vehicle. 1843. Earl of Dundonald fitted vertical, tubular boilers to U.S. steamships *Atlantic* and *Pacific*. 1844. Lancashire boiler *inv.* Sir William Fairbairn. 1866. Oil-fired boiler *inv.* C. J. Richardson, of Woolwich. 1867. Tubular boiler *inv.* Messrs. Babcock and Wilcox (U.S.). 1875. Solar-heated boiler *inv.* Mouchot (Fr). (*See also* Solar power.)

BOLLÉE, Amédée (Fr) (1844–1917) 1873. *Des.* steam road vehicle. (*See* Road vehicles, Steam.)

BOOTH, Henry (1788–1869) 1828. *Inv.* locomotive-type steam boiler.

Bolometer 1880. *Inv.* by Samuel Pierpoint Langley (1834–1906).

Bolting, Dusting and Sifting Machines 1623. John Rathbone *inv.* bolting machine. 1686. John Finch, John Newcomb and James Butler *inv.* wire screen sifting machine. 1765. John Milne *des.* sifting machine or dresser to work by wind or water. 1775. George Robinson *des.* man- or horse-operated dressing machines for grain. 1783. Benjamin Blackmore manufactured bolting-cloth for bolting machines.

Bomb *Inv. c.* 1495. 1522. Used at Siege of Rhodes. Re-*inv.* 1835. J. M. Fieschi (1790–1836).

Bomb, Hydrogen (H. bomb) 1951 (May). First test of in New Mexico, U.S., 1952. (Nov 1). 65-ton U.S. bomb dropped on Eniwetok Atoll, Pacific Ocean. 1953. (Aug 4). Russians explode experimental bomb. 1954 (Mar. 1). Second hydrogen bomb dropped on Eniwetok Atoll. 1954 (Apr. 6). Third hydrogen bomb dropped on Marshall Island.

Bones, Napier's 1617. *Inv.* John N. Napier (1550–1617).

Bone-setting 1620. First scientifically practised.

Book 184 B.C. Attalus (*Leg*) *inv.* first book with leaves of vellum. *c.* 1440. First book printed with movable type (Ger). 1448. Johann Gensfleisch Gutenberg (Ger) (*c.* 1397–1468) printed first book from cast type at Mainz. (? 1454 by Coster, of Haarlem, Holland.)

Bookcase, Revolving *c.* A.D. 823. Octagonal revolving bookcase with braking device *inv.* in China.

Boots *c.* 907 B.C. First mentioned by Homer.

Bordeaux Mixture (insecticide) 1885. *Inv.* Dr. Gayon (Fr) of Bordeaux.

Boring Machines 1540. First mention of boring machine (for cannon) by Beringuccio (It) (1480–1539). 1769 John Smeaton (1724–92) *inv.* metal boring machine. 1775. John Wilkinson (1728–1808) *inv.* metal boring machine and erected one in Denbighshire. 1799. William Murdock (1754–1839) *inv.* worm-driven cylinder boring machine. Also *inv.*, 1810 circular crown-saw boring machine.

Boron (element) 1809. *Disc.* Joseph-Louis Gay-Lussac (Fr) (1778–1850) and Louis Jacques Thénard (Fr) (1777–1857).

BÖSCH, Robert (Ger) (1874–1940) *c.* 1890. *Inv.* internal combustion engine L.T. magneto ignition system and H.T. magneto in 1902.

BÖTTGER, Johann Friedrich (Ger) (1685–1719) 1705. Made first hard porcelain in Europe at Meissen, Dresden.

Bottle-making Machine, Automatic 1899. *Inv.* Michael Joseph Owens (U.S.) (1859–1923), with E. D. Libby (U.S.). 1920. Modern machine evolved.

Bottling (preserving) 1840. Fish and fruit first successfully bottled.

BOULTON, Matthew (1728–1809) 1787. *Inv.* the Venturi tube.

BOURDON, Édouard (Fr) (1808–84) 1849. *Inv.* pressure-gauge bearing his name.

BOUSSINGAULT, Jean-Baptiste (Fr) (1802–87) *c.* 1850. *Disc.* that plants absorbed nitrogen from the soil and their carbon from the atmosphere (the nitrogen cycle).

BOUTON, Georges (Fr) (1847–1938) 1883–97. *Des.* and *const.* many steam road vehicles (*q.v.*).

Bow and Arrow 30000–15000 B.C. First clear representations of bow and arrow from Sahara Desert area of North Africa.

Bowden Cable (control) 1912. *Pat.* by G. F. Larkin and the Bowden Wire Company for use in motor-cycle and car controls.

BOYDELL, John 1846. *Pat.* track-laying traction-engine (*q.v.*).

BOYLE, Robert (Irish) (1627–91) 1662. *Disc.* that volume of gas varies in inverse proportion to its pressure — Boyle's Law. *See also* Henri Marriotte (Fr).

Boyle's Law (Marriotte's Law) 1662. *Disc.* Robert Boyle, *see* previous entry.

BOYS, C. V. (1855–1944) 1912. *Des.* solar power plant.

BOZEK, Joseph (Aus) 1815. *Des.* and *const.* a steam road vehicle carrying three persons.

Brace, Hand 1424. First *des.* 1420. Shown on panel of Muster Franck's St. Thomas's altar. (First example of compound crank.) 1460. Reproduction in book by Jean de Tavernier d' Audenarde (Fr).

Braces 1820. First made of india-rubber by J. Hancock.

Brackets, Poisson's 1925. Adrien Maurice Dirac (Fr) *disc.* their significance in mathematics. *Inv.* by Simon Denis Poisson (Fr) (1781–1840).

BRADLEY, James (1692–1762) 1727. *Disc.* aberration of light of fixed stars.

BRAHE, Tycho (Den) (1564–1601) Astronomer. Regarded the earth as the centre of the solar system—the "Tychonic System."

Brain, Human Speech centre of human brain *disc.* by Paul Broca (Fr) (1824–80), of Paris.

Brake, Air 1844. James Nasmyth first *pat.* for air brake. 1780. Safety air brake fitted to treadmill-operated crane *des.* by Pinchbeck. 1879. John A. P. Aspinall *inv.* effective vacuum brake. 1875. George Westinghouse *inv.* continuous compressed air brake. 1869. Westinghouse *inv.* "straight" air brake.

Brake, Steam 1833. *Inv.* Robert Stephenson.

Brake, Hydraulic 1877. *Inv.* for trains by Barker. Also by Clarke. 1904. *Inv.* for motor cars by F. G. Heath. 1917. Lockheed hydraulic brake for motor cars by M. Loughead.

Brake, Motor car front wheel 1904. *Inv.* P. L. Renouf. 1909. Front wheel brakes fitted to Argyll motor car. 1906. Front wheel brake *inv.* Allen (an *imp.* on the Renouf patent).

Brake Linings (automobile engineering) 1900. *Inv.* Herbert Frood.

Brakes, Disc 1906. Bronze disc brakes installed on Darracq motor cars. 1913. Disc brakes installed on A.C. motor cars.

Brake, Railway train Continuous brake and train signal *inv.* by James Newall, of Bury, Lancs.

Brake, Servo 1919–26 Dewandre brakes *pat.* A. Dewandre, F. L. Goodyear and J. G. P. Thomas. 1924–29. Duo-servo brakes *pat.* V. Bendix.

Brake Shoes 1902. Internal-expanding, cam-operated brake shoes *inv.* Louis Renault. 1917. Adjustable brake shoes *inv.* R. Stevens.

Brake, Water *c.* 1880. *Inv.* by William Froude.

BRAMAH, Joseph (1749–1814) 1796. *Inv.* hydraulic press. 1784. The "Bramah" lock. 1778. Water-closet, etc.

BRANCA (*c.* 1629) *Desc.* many machines of his day in his book *Machines*.

Brandy 1667. First used. 1748. First distilled from potatoes.

BRAY, William 1856. *Inv.* steam traction-engine with wheels fitted with optional spikes to increase adhesion on soft ground.

Bread 170 B.C. First made by regular bakers of Ancient Rome.

Bread, "Aërated" 1856. *Inv.* Dr. Dauglish.

BREWSTER, Sir David (1781–1868) 1833. *Disc.* fluorescence in chlorophyll. 1838. *Disc.* fluorescence in fluorspar.

Brickmaking Machinery First invented, 1839.

Bridge *c.* 4000 B.C. Earliest example of unburnt brick arch still standing at Nippur, Chaldea. A.D. 450. King Chosroes I built 80 ft. wide, 90 ft. high arch at Ctesiphon. 104. Apollodorus of Damascus built 3,000 ft. timber bridge on stone piers over the River Danube for Emperor Trajan. *c.* 7th cent. Earliest example of regular stone-voussired arch at the Phoenician-built canal of the Marta, near Graviscae. 860 Y-shaped bridge built at Croyland, Lincolnshire. 994. William of Malmesbury built bridge over River Thames at London. 1176. First stone bridge over River Thames at London *des.* by Peter of Colechurch or Colchester. 1187. St. Bénezèt *des.* and built famous Pont d'Avignon over River Rhône (20, 100 ft. arches). 1570. First known *des.* for truss bridges evolved by Andrea Palladio (1518–80). 1697. First stone bridge in U.S. built at Pennepecka, Germanstown, Pa. 1755. 3-arched cast-iron bridge over River Rhône at Lyons *des.* by French army engineer Garvin. (Only one arch completed owing to high cost.) 1755. Brothers Grubenmann built large timber-span bridge over River Rhine near Schaffenhausen. 1758. A similar bridge over River Lim-

mat near Zurich erected, and another at Wittengen. 1787. First cast-iron bridge in England *des.* and erected by Abraham Darby III and foundry-owner John Wilkinson, over River Severn at Colebrookdale. 1760. Scotsman Robert Myln *intro.* multiple wedge centreing on Blackfriars bridge over River Thames at London. 1800. James Finley (U.S.) *des.* first suspension bridge in U.S. with iron chains and rods. 1801. Finley *des.* level-floored suspension bridge of modern type, across Jacob's Creek, Westmoreland County, Pa. 1812. 340 ft. timber bridge built across Schuykill River, U.S. 1815–25. Thomas Telford (1757–1834) *des.* and built 580 ft. span flat-link, chain suspension bridge carrying Holyhead Turnpike across the Menai Straits to Isle of Anglesea. (The chain-link system was *inv.* by Samuel Brown.) 1820. Ithiel Town (U.S.) *pat.* wooden-beam lattice truss bridge. 1825. Wire suspension bridge erected at Lyons by Marc Séguin. 1829. Carrolton railway viaduct *des.* and built over Gwynne's Falls, Baltimore, Maryland, U.S. by James Lloyd. 1835. First cast-iron bridge in U.S. *des.* by John Herbertson and built by John Snowden over Dunlap's Creek, Brownsville, Pa. 1840. William Howe (U.S.) *inv.* wrought-iron (later, steel) bridge truss. 1840. Andrew Thomson, of Glasgow *des.* and *const.* bridge of hollow girders of boiler-plates to carry road over Pollock and Govan Railway. 1840. Earl Turnbull builds first cast-iron girder bridge in U.S. over Erie Canal at Frankfurt, N.Y. 1841. John Augustus Roebling *inv.* parallel, instead of twisted wires for suspension bridge *const.* 1844. Tubular bridge girder *pat.* France by Dr. Jules Guyot. (*Pat.* England 1846.) 1844. Brothers Caleb and Thomas Pratt (U.S.) *pat.* design for special truss for steel bridges. 1846. Wrought-iron girders first made by riveting by shipbuilder William Fairbairn. *c.* 1850. Britannia tubular bridge across Menai Straits *des.* Robert Stephenson in collaboration with William Fairbairn and Eaton Hodgkinson. 1852. J. A. Roebling commences 822 ft. span suspension bridge over Niagara River, U.S. 1867. Heinrich Gerber (Ger) *des.* first modern-type cantilever bridge over River Main. 1868. Finck (U.S.) *inv.* special form of bridge truss. 1873. Wood and steel truss bridge *inv.* T. W. Pratt (U.S.). 1874. James Buchanan Eads (U.S.) erects first steel arch bridge in U.S. at St. Louis, Mo. 1876. C. Shaler Smith *des.* first cantilever bridge in U.S.—over Kentucky River. 1895. William Scherzer of Chicago *inv.* rolling-lift bridge and erects the first of this type in his home city. 1898. Françiose Hennebique (Fr) *des.* and erects 172 ft. span reinforced concrete (*q.v.*) bridge at Châtellerault. 1905. First reinforced concrete cantilever bridge in U.S. built at Marion, Iowa (3 × 50 ft. spans).

Bridge, Pontoon (floating) 1st cent. B.C. Caesar used pontoon bridge to cross River Rhine. *c.* 490 B.C. Xerxes built pontoon bridge across the Hellespont at Abydos. 1710. Pontoon bridge *inv.* (?) by Françoise Joseph Camus (Fr). 1773. D'Herman (Fr) *inv.* pontoon bridge. 1804. First pontoon bridge in U.S. across Collin's Pond, Lynn, Mass. 1655. Portable bridge mentioned in Marquis of Worcester's *Century of Inventions No. 28.* 1944. Portable bridge *inv.* Sir Donald Bailey.

Bridge, Wheatstone (electric) *c.* 1820. *Inv.* credited by Charles Wheatstone (1802–75) to Christie.

Bright's (kidney) Disease 1837. *Disc.* and named by Dr. R. Bright (1789–1858) of Bristol and Guy's Hospital, London.

Britannia Metal First made at Sheffield, 1770.

Broad Arrow 1698. First used to mark dockyard and naval property.

Broadcloth First made in England, 1197.

Brocade First made at Lyons, France, 1757.

Bromine (element) 1826. *Disc.* in salt water by Antoine Jérôme Balard (1802–76).

BROTHERHOOD, Peter (1838–1902) 1871. *Des.* three-cylinder steam-engine.

Brougham (horse-drawn vehicle) 1839. *Inv.* Baron Henry Peter Brougham (1778–1868).

Brownian Movement 1827. *Disc.* by botanist Robert Brown in non-living matter. 1905. Brownian movement explained by Albert Einstein (1879–1955).

BRUNO, Giordano (It) (1473–1543) 1584. First to represent the universe as infinite. (*See also* William Gilbert, Copernicus, Galileo and Kepler.)

BRUSH, Charles Francis (U.S.) (1849–1929) 1876. Designed his first dynamo and the following year, his first arclamp.

Brushes *c.* 1400 B.C. Earliest record of brushes—Egypt.

Bucklers (armour) 1370 B.C. *Inv.* Proteus and Achrisius of Argos. *See also* Cuirass.

Bude Light *See* Lime light.

BUFFON, Georges-Louis Leclerc, Comte de (Fr) (1707–88) *See* Photometry, Optics, and Burning-mirror.

Bulkhead, Watertight Used in Ancient China. 1795. *Pat.* Samuel Bentham. Pre-1798. *Intro.* Capt. Schanks into G.B. on experimental ship "Trial." Mentioned in Marquis of Worcester's *Century of Inventions, No. 12.*

Bullet (arms) 1514. Stone bullets mentioned. 1550. Iron bullets mentioned. Pre-1700. Lead bullets mentioned. 1833. Conchoidal cup rifle bullet *inv.* Capt. Minié (Fr). 1853 Minié's *inv.* modified by Pritchett.

Bullet, Explosive (arms) 1495. Evolved by Leonardo da Vinci. 1573. First practical explosive bullet. 1602. First efficient explosive bullet *inv.* Renard Ville (Fr).

Bullet, Incendiary 1759 Torre (Fr) *inv.* incendiary bullet. 1772. Wooden gun which fired arrows which burst into flame when they struck *inv.* J. P. Costé (Fr), of Dauphinais. 1786. Fabre (Fr) *inv.* "Bellegarde" bullet.

Bumpers, Motor car 1905. *Inv.* Frederick R. Sims.

Buna S (artificial rubber) 1933. First *prod.* by F. R. Bock. 1939. Commercial production achieved.

BUNSEN, Robert Wilhelm (1811–99) (Ger) 1861. With G. R. Kirchoff *des.* an improved spectroscope, also *inv.* the shadow photometer.

Buoy 1866. First electrically lighted buoy *inv.* Adolphe Miroude (Fr).

BURALI-FORT (It) 1897. Proposed theory of antinomies of aggregates.

Burglar Alarm 1655. Mentioned by Marquis of Worcester in *Century of Inventions, No. 72.*

Burning Mirror 214 B.C. *Inv.* by Archimedes and used to fire ships of Marcellus's fleet at Syracuse. 1739. Archimedes's success proved by Buffon.

BUSCHMANN (Ger) 1821. *Inv.* mouth-organ. 1822. *Inv.* accordion.

Butylene (chemistry) 1825. *Disc.* Michael Faraday.

Buys Ballot's Law (meteorology) Enunciated *c.* 1850 by Buys Ballot (1817–90), of Utrecht, Holland.

C

Cab, Hansom 1833. *Inv.* Joseph A. Hansom.

Cab (Cabriolet) 1823. *Intro.* into London.

Cables, Iron 1808. *Intro.* for ship's rigging. 1816. Chain cables used at bombardment of pirates at Algiers. 1840. Chain cables commonly used in ship's rigging. *See* Rope, Rope, wire, Cable, chain, Chain.

Cable, Oil-filled electric 1914. High-tension type *inv.* 1914, Emanuelli. *c.* 1925. 132,000 volt cable manufactured.

Cable, Submarine telegraph 1850. First laid, Dover to Cap Grinez. 1858. First Atlantic submarine cable laid. (Lasted in use for only 3 weeks.)

Cable, Submarine telephone 1912. Submarine telephone cable laid across Straits of Dover. 1956. Submarine telephone cable laid Nova Scotia to Newfoundland.

Cadmium (element) 1817. *Disc.* Stromeyer (Ger).

Cæsium (element) 1860. *Disc.* Robert Wilhelm Bunsen (Ger) (1811–99).

CAGNIARD-DE-LA-TOUR, Charles (Fr) (1777–1859) 1819. *Inv.* the siren. Pioneered work on the propagation of acoustic waves in water.

CAILLETET, Louis Paul (Fr) (1832–1913) *c.* 1877. Pion-eer of early experiments in the cryogenic field.

Caisson, Compressed air 1790. *Inv.* Perronet (Fr). 1830. First *intro.* for foundation construction by Sir Thomas Cochrane, later the Earl of Dundonald.

Calcium (element) 1808. *Disc.* Sir Humphrey Davy (1778–1829).

Calcium Carbide 1891. First *prod.* in electric furnace by Börchers. (*See also* Acetylene.)

Calculating Machines *See* Computers.

Calculus 1674. *Inv.* Gottfried Wilhelm Leibnitz (Ger) (1646–1716) and made public in 1684.

Calculus, Differential 1665. Sir Isaac Newton (1642–1727). *Inv.* "fluxions." 1677. G. F. R. Leibnitz (1646–1716) *inv.* differential calculus.

Calculus, Exponential *c.* 1720. *Disc.* Jean Bernouilli.

Calculus, Infinitesimal 1615. Johann Kepler (Ger) (1571–1630) *pub.* "Nova Stereometria" containing the germs of I.C. idea.

Calculus of Matrices 1858. *Inv.* A. Cayley, mathematician.

Calculus of Variations 1767. *Disc.* Joseph Louis Lagrange (Fr) (1736–1813). Theory *prop.* by Adrien Marie Legendre (Fr) and further *dev.* by K. G. Jacobi, K. F. Gauss,

Simeon Denis Poisson, and Augustin-Louis Cauchy.

Calculus, Tensor *Inv.* M. M. Gregorio Ricci (1853–1925) and Tullio Levi-Civita (It) mathematicians of Padua University.

CALDANI, Leopoldo Marc-Antonio (It) 1756. *Obs.* contractions of a frog preparation under the influence of electricity. Galvani's name has become attached to the complex phenomena.

Calendar 3500 B.C. Egyptians used a reasonably accurate calendar. 3000 B.C. Egyptians *intro.* civil year of 365 days divided into 12 months of 30 days. 2500 B.C. 12-month lunar calendar *intro.* into Egypt. 357 B.C. Lunar calendar revised in Egypt. 239 B.C. Egyptians realized and corrected "drift" of calendar and *intro.* leapyear. 45 B.C. Julian calendar devised by Sosigenes (54 B.C.–A.D. 24) and *intro.* by Julius Caesar. 1582. Gregorian calendar reform: Oct. 5 became Oct. 15. 1752. Gregorian calendar adopted in Britain. 1918. Adopted in Russia. 1927. Adopted in Turkey.

Calico 1631. *Intro.* into England by East India Company.

Calico Printing *See* Printing, Textile.

Californium (element) 1950. *Disc.* Glen T. Seaborg (U.S.) and A. G. Liorso (U.S.).

Calipers A.D. 79. Bronze outside calipers used in Pompeii. 1540. Internal calipers *inv.* by an artificer of Nuremberg. 1851. Vernier calipers *inv.* J. R. Brown (U.S.).

Calotype Process 1841. *Inv.*

bu W. H. Fox-Talbot. *See* Photography.

Calorimeter *Inv.* Benjamin Thompson (Count Rumford) (U.S.) (1753–1814). 1870. Ice calorimeter *inv.* Robert Wilhelm Bunsen (Ger) (1811–99) (also attributed to Lavoisier). Bomb calorimeter *inv.* P. E. M. Berthelot (Fr).

Calorimeter (medicine) Respiration calorimeter *dev.* by Atwater, Rose, and Benedict.

Cam 983. First useful application in the West in fulling mill on River Serchio, in Tuscany. 1010. Water-driven, cam-operated trip-hammers in use at Schmidmülen, in the Oberpfalz.

Camera *See* Photography.

Camera, Electron 1931. Electron diffraction camera evolved by G. I. Finch. (*See also* Electron microscope.)

Camera Lucida *Inv.* William Hyde Wollaston (1766–1828).

Camera Obscura. 1100. Mentioned by Alhacin (Arab). *c.* 1500. Mentioned by Leonardo da Vinci (It). (1452–1519). *c.* 1600. Convex lens substituted for pin-hole by D. Basbaro and Giovanni Battista Porta (It) (1543–1615). 1292. *Inv. att.* to Roger Bacon (*c.* 1210–93). Rainer Gemma Frisius (Hol) used camera obscura to *obs.* solar eclipse on Jan. 25, 1544. Johann Kepler (1571–1630) suggested use of convex and concave lens to increase size of image. (*See also* Photography.)

Canals 1209. Naviglio Grande, irrigation canal in Italy opened. Later used to transport marble from Lake Maggiore to build Milan cathedral. 1391. Elbe-

Stecknitz watershed surmounted by a canal with two locks with a rise of 16 ft. and followed by a 12 ft. deep canal from Lübeck to Hamburg. 1458. 12 mile canal with 180 ft. descent by 18 locks built, the mitred lock-gates being *inv.* by Leonardo da Vinci.

Candle A.D. 1. Huge candle of date found at Vaison, near Orange. Now in British Museum. A.D. 50. Candle-making apparatus found in ruins of Herculaneum. 1760. Wax for candles bleached in the sun at Field's London Candle Works. 1832. De Milly commenced making candles at Barriére de l'Etoile, Paris.

Candlesticks First mentioned (*c.* 1491 B.C.) Exodus xxxvii, 17.

Canning Process 1795. French confectioner François Appert *inv.* process for food preservation (by bottling); building a plant for the purpose at Bermondsey, London. *See also* Bottling.

CANNIZARRO, Stanislao (1826–1910) 1858. Pioneered system of a consistent system of atomic weights.

Cannon 4th cent. A.D. Vegetius mentions use of scorpiones, arcubalistae, fustibuli, fundae, ballistae and movable towers. 1259. Chinese made bamboo cannon. 1275. Chinese made metal cannon. 1324. Cannon used at siege of Metz. 1328. Earliest record of use of guns during invasion of Scotland. 1343. Cannon *intro.* into war operations by Moors against Alonzo, King of Castile. 1346. Cannon used by Edward III against the French at battle of Crecy. 1370. Cast-iron mortar *intro.* by Merklin Gast, of Augsburg. Late 14th cent. Wrought-iron cannon firing small iron shot and stone balls weighing up to 450 lb. in common use. 1382. 13 ton cannon (bombard) 16 ft. long and 25 in. bore built of welded longitudinal iron strips at Ghent. Its granite ball shot weighed 700 lb. 1404. 4½ ton cannon cast in Austria. 1494. Cannon firing range for battering purposes was about 100 yd. 15th cent. Trunnion mounted cannon *intro.* 1540. Cannon rifling *intro.* by A. Küttner. 1574. Largest English cannon 4 tons weight, 8¾ in. bore, used 60 lb. shot. 1543. Cannon cast at Buxtead, Sussex. Late 16th cent. Hollow shot filled with explosive or incendiary mixtures fired from mortar cannon. 1580. Petard *intro.* 1586. Cannon made in England at Long Ditton. 1697. Howitzer *intro.* 1720. First cannon cast in one piece by Keller, of Cassel. 1799. Carronade *inv.* by Gascoigne, director of Carron Ironworks, Scotland. 1859. Rifled bronze cannon used at battles of Magenta and Solferino. 1876–78. Armstrong rifled cannon *dev.* by Sir William Armstrong. Alfred Krupp (Ger) (1810–87) *dev.* cast-steel cannon.

Cannon-boring Machine 1714. Jean Moritz (Swit) *inv.* horizontal drill for boring cannon. 1716. Water-driven drills for boring small arms *inv.* Villons (Fr). 1774. John Wilkinson (1728–1808) *inv.* cannon boring machine.

Cannon, Breech-loading 1715. *Inv.* by De la Chaumette (Fr).

Cannon, Steam 1803. William Murdock (1754–1839) fired a 1 in. ball from a steam cannon at Soho Works, Birmingham. *c.* 1824. Steam cannon *dev.* by Perkin. (Leonardo da Vinci (1452–1519) sketched idea for steam cannon and Cesariano pictures steam-exploded grenades in 1521.)

CANTON, John (1718–72) One of the English pioneers of static electrical machinery.

Capillaries (anatomy) *Disc.* Marcellus Malpighi (It) (1628–94).

Capillary Attraction 1600–35. First *obs.* by Niccolo Aggiunti of Pisa. 1805. Thomas Young (1773–1829) formulated theory of capillary attraction. Another theory put forward by A. C. Clairaut (Fr).

Capstan, Ship's Late 15th cent. Mentioned by Columbus on his second voyage as "cabestante." Re-*inv.* and *imp.* by Sir Samuel Morland (1625–95).

Carbohydrate 1862. Term first used by Ernst Wagner (Ger).

Carbon (elment) 1788. Proved a distinct element by Antoine-Laurent Lavoisier (Fr) (1743–94).

Carbon dioxide gas First recognized by Jan Baptista van Helmont (Hol) (1577–1644) 1855. First liquefied and solidified by Thirolier (Fr). *See also* Mineral waters and Refrigeration.

Carbon Disulphide 1796 *Disc.* Wilhelm August Lampadius (Ger) (1772–1842).

Carbon Monoxide Gas 1776. *Disc.* by Joseph M. F. de Lassonne (Fr) (1717–88) and Joseph Priestly (1733–1804) independently.

Carborundum (silicon carbide) 1891. Process for making evolved by Edward Goodrich Acheson, of Monongahela City, Pa., U.S.A.

Carburettor (internal combustion engine) 1893. Float-feed, spray carburettor re-*inv.* Maybach (Ger) (1846–1929). 1896. Surface carburettor *inv.* Petréano. Wick carburettor. *c.* 1897. Various types *inv.* W. G. Buck, F. Lanchester, and by *invs.* of "Papillon and Balbi" carburettors. 1904. "S.U." carburettor *inv.* by G. H. and T. C. Skinner. 1906. "Claudel-Hobson carburettor *inv.* C. H. Claudel. 1907. "Zenith" carburettor *inv.* F. Bavery (Fr). 1866. *Inv.* by Bowditch. 1888. Edward Butler *des.* and *pat.* a spray carburettor.

CARDAN, Jerome (Jeromy Cardanus) (It) (1501–1576) Physician. 1551. *Inv.* universal joint, or Cardan-shaft for driving shafts. (*See* his book, *De Subtilitate*).

Cards, Playing 1390. First known card game, piquet, *inv.* 1391, said to have been *inv.* in France to amuse King Charles IV.

Carding Machines (textiles) 1748. Lewis Paul *inv.* and *pat.* two carding machines. 1782. Antoine Germondy, of Lyons *pat.* hand or water-operated carding machines, *imp.* by Fournier de Granges (Fr) 1784. 1784. Mule-powered carding machine *inv.* Simon Pla (Fr).

1791. Carding machine for wool and felting *inv.* Serrazin (Fr). 1796. Carding machine for wool *pat.* by Louis Martin, and *imp.* 1797 by Tellié (Fr). 1792. Paris physician Lhomonde *inv.* carding machine for cotton. 1823. Differential roving winding-gear *intro.* by Mansfield tinsmith and used by Henry Houldsworth of Glasgow on his carding machine of 1825. 1822. Differential carding machine *inv.* Asa Arnold (U.S.). (Model brought to England same year as Houldsworth's *pat.* was taken out.) 1790. Carding machine (comber) *inv.* Rev. Edmund Cartwright. 1851. Carding machine *inv.* Loshua Heilmann (Ger).

Card-making Machine (textiles) 1777. *Inv.* Oliver Evans (U.S.) (1755–1819).

Caricature Drawing 1320. Style *inv.* Bonamico Buffalmacco (It) (1262–1340).

CARLISLE, Sir Anthony (1768–1840) 1800. With William Nicholson first decomposed water into oxygen and hydrogen with aid of an electric pile.

Carlsbad Mineral Springs 1370. *Disc.* Emperor Charles IV.

CARNOT, Lazare-Nicholas-Marguerite (Fr) (1753–1823) 1792. One of the pioneers of the fundamental principles of mechanics and the theory of the study of motion.

CARNOT, Nicholas-Léonard-Sadi (Fr) (1796–1832) 1824. *Intro.* the idea of the heat cyclic process now bearing his name.

Carnot Cycle *Prop.* 1824.

N. L. S. Carnot (Fr). *See* above.

Carpet 1589. *Intro.* into France. 1744. *Intro.* into England.

Carriage, Road *c.* 1486 B.C. *Inv.* ascribed to Erichthonius of Athens.

Carriage, Sail 552. Used in China, constituting the first example of high-speed travel.

Cart Pre-3000 B.C. Two and four-wheeled carts used in Syria.

Cartesian Co-ordinates (biology) 1637. *Intro.* by Réne Descartes (Fr) (1596–1650).

Cartwright, Rev. Edmund (1743–1823) Vicar of Brampton, near Chesterfield. 1785. *Inv.* Power loom.

Carving machine Wood Pre-1846. Carving machine for ivory *inv.* Cheverton. 1846. Carving machine for wood *inv.* T. B. Jordan.

Casein 1780. *Disc.* by C. W. Scheele (1742–86). 1919. Used as ingredient of plastics and some nylon yarns.

CASSINI, Giovanni Domenico (It) (1625–1712) With C. Huygens, and Giacomo Filippo Maraldi, one of the first astronomers to measure the period of axial rotation of the planet Mars.

Casting, Continuous method of 1925. Byron E. Eldred (U.S.) *inv.* continuous casting process for non-ferrous metals. 1927. Dr. Siegfried Junghans *inv.* the Rossi-Junghans continuous casting process for non-ferrous metals. 1949 (Mar.). Dr. Junghans applies his process to steel.

Casting, Plastic Metal Shell 1941. *Dev.* Johannes Croning (Ger), of Hamburg.

Catalysis (chemical) 1806. First *desc*. Charles Bernard Désormes (Fr) (1777–1862) and Nicholas Clément (Fr) (1779–1841).

Catalytic Action *Disc*. J. J. Berzelius (1779–1848), who named catalysts.

Catalytic Force *Disc*. by L. J. Thenard (1777–1857). 1825. Catalytic force used to ignite gas by Döbereiner.

Catapault 399 B.C. Said to have been *inv*. by Dionysius of Syracuse. 397 B.C. Stone and javelin catapaults used against Carthaginians. 287–212 B.C. *Inv*. (?) by Archimedes and used with great force against Romans besieging Syracuse. 332 B.C. Torsion catapault used at siege of Troy. (Later Roman catapaults could throw up to 500 yd. and ballistas (crossbows) over that distance. 1139. Lateran Council under Pope Innocent II prohibited use of catapaults except against infidels.

Cataract (surgery) Late 16th cent. Georg Bartisch devised means of extracting cataract from eye. Middle Ages. Cataract operation performed by Arabian physicians. 1706. Antoine Mâitre-Jean explained nature of cataract. 1709. Michael Brisseau, of Tournai independently explained cataract. 1730. Françoise Son Étienne du Petit (Fr) removed cataracts with some success.

Catheter, Eustacian (surgery) *Inv*. Guyot (Fr), at Versailles.

Cathode Rays 1859. *Disc*. Plücker. *See also* X-rays.

Cathode Rays, Charge-to-mass ratio of 1897. First measured by J. J. Thomson and Weichert.

CAUCHY, Augustin-Louis (Fr) (1789–1857) Mathematician who helped to perfect the undulatory theory of light and established a formula expressing the speed of light in a medium, and thus the refractive index of the medium as a function of the wavelength.

CAVALIERI, Bonaventura (It) (1598–1647). *c*. 1660. With Niccolo Zucchi, Marin Mersenne, Rene Descartes and James Gregory, anticipated Sir Isaac Newton in making a reflecting telescope.

CAVENDISH, Henry (1731–1810) Produced "fixed air" (carbon dioxide) from marble and acids; also produced "inflammable air" (hydrogen) Proved that water was a compound of oxygen and hydrogen by passing sparks through a mixture of the two gases and obtaining water as dew.

CAXTON, William (1412–92) 1474. Printed first book in England. *The Game and Play of Chesse*.

CAYLEY, Sir George (1774–1857) Father of British aeronautics. 1843. Built steam-powered aeroplane. 1896. *Des*. model helicopter. 1825. *Pat*. a track-laying tractor. *See also* Henson, Stringfellow and Maxim.

Celanese (rayon acetate) 1865. Process reactions to make rayon acetate known to Schützenberger (Ger). 1904. Commercial production commenced. *See also* Rayon.

Cell (biology) 1839. Cell theory of tissues *prop.* by Theodore Schwann (Ger) (1810–82). 1665. Word "cellular" used by Robert Hooke (1635–1703) in his book *Micrographica*.

Cellophane (Cellon) 1912. First *prod.* by Jacques Edwin Brandenberger (Swit). 1869. *Prod.* from cellulose acetate (*q.v.*). 1926. Moisture-proof Cellophane *inv.* William Hale Church and Karl Edwin Pringle, of Du Ponts, U.S. 1933. First commercial *prod.* of Cellophane commenced in U.S.

Celluloid 1860s. *Prod.* from camphor and pyroxylin (cellulose nitrate) by Brothers. John Wesley and Isiah Hyatt, of Albany, N.Y., U.S.

Cellulose 1862. "Parkesine" (cellulose) made from guncotton by Alexander Parkes, of Birmingham. It could be made to simulate ivory, tortoiseshell, wood, or india-rubber; and was an excellent electrical insulator.

CELSIUS, Anders (Swed) (1701–44). 1742. *Intro.* centigrade thermometer scale with freezing and boiling-points of water as fixed points.

Cement *c.* 150 B.C. Lime mortar made by Romans at Pozzuoli. 1796. James Parker re-*disc.* puzzolina hydraulic cement. John Smeaton (1724–92) *pat.* "Roman" (hydraulic) cement. 1824. L. J. Vicat (Fr) *inv.* cement made from chalk and clay. 1824. "Portland" cement re-*inv.* at Wakefield by William Aspdin (1779–1855). 1839. Vicat's cement used in *const.* of Cherbourg harbour works. 1844.

I. C. Johnson made artificial hydraulic cement.

Census 1490 B.C. First census taken by Moses. 1017 B.C. Census taken by King David. 561 B.C. First Roman census taken by Servius Tullius. 1801. First British census.

Central Heating 1777. First modern system *inv.* Bonnemain (Fr), who installed it at Château du Pecq, near Germain-en-Laye.

Centrifuge 1860. Napier used centrifuge in *inv.* of continuous-action sugar-drying machine. 1890. C. G. P. de Laval (1845–1913) *inv.* centrifuge cream separator. (*See also* Separators.) 1843. Revolving-bucket type *inv.* by Lawrence Hardman for sugar-drying. (Also *pat.* by Seyrig (Fr).) 1850. Self-centring centrifuge *inv.* by Sir Henry Bessemer and simultaneously, but independently by David McCoy Weston (U.S.). 1849. Centrifuge to treat sugar-loaves *inv.* by Broonan.

Centrosome (cytology) 1888. Name *intro.* by Theodore Boveri (1862–1915).

Ceramic, Electrical 1941. High di-electric constant of barium-titanic oxide ceramics *disc.* (U.S. *Pat.* 2429588.)

Ceres (planet) 1801 (Jan. 1). *Disc.* Guiseppe Piazzi (It) (1746–1826), at Palermo.

Cerium (element) 1803. *Disc.* Martin Klaproth (Ger) and independently by J. Berzelius and William Hisinger (both Swed).

CESSART, Louis Alexander de (Fr) (1719–1806) 1787. *Des.* and used on French roads horse-drawn, cast-iron rollers.

Cetaceans First recognized as mammals by Pierre Belon (Fr). **CHADWICK, James** (*b.* 1891–) Atomic physicist. Transmuted many lighter elements with Sir Ernest Rutherford, in 1919.

Chains (cable) 2500 B.C. Chains made of rings folded in half and passed through next link found in cemetery at Ur of the Chaldees. 8th cent. B.C. Chains with S-shaped links of cast bronze found at Nimroud. 57 B.C. Iron chain cables used by Venetii of Brittany. 1771. Chain cables suggested in lieu of hempen cables, but not used until 1808 by naval surgeon Slater. 1811. Chain cables *intro.* Capt. Brown on wrought-iron ship *Penelope*. Twisted links were used. 1822. Sowerby *pat.* inward-bent link. 1825. Brunton *intro.* straight-link and cross-rod chain. 1828. Hawks *pat.* elliptical, thick-ended link. Jacques de Vaucanson (Fr) (1709–82) *inv.* the "Chaine de Galle" with eye-shaped links.

Chains, Driving 1588. Wire-link type mentioned by Agostino Ramelli (1531–90) in his book *Le Diverse et Artificiose Machine.* 1852. Andrew Crestadoro *inv.* endless chain roadway. 1856. First English chain *pat.* Elias Robison Hancock, for chain and chain-wheel. 1860. John Fowler, Jnr. and William Worby *inv.* driving chains for agricultural tractors. 1861. Driving chains with side guard links *inv.* Thomas Green and Robert Mathers to prevent chains slipping off wheels. 1863. Richard Hornsby and John Bonnall *pat.* driving chain with

rollers on links. 1863. Thomas Sturgeon *inv.* T-link driving chain. 1864. James Lancelot *pat.* driving chain of corkscrew shape with wheels to match. 1864. James Slater *inv.* toothed driving chains for working with toothed cog-wheels. 1866. John Erskine Brown *pat.* driving chains and chain-wheels. 1864. Roller driving chain *inv.* Slater. 1868. Roller pitch driving chain *inv.* Slater. *c.* 1870. Brampton *imp.* Slater's driving chain. 1879. Hans Reynold started business and *imp.* Slater's and Brampton's driving chains. 1879. James Starley built first bicycle with driving chain. 1880. Hans Reynold *inv.* silent driving chain. 1546. Vaucanson-type *illus.* by Agricola. 1750. French wire-link driving chain *inv.* by Jacques de Vaucanson (Fr) (1709–82).

Chain Shot 1664. *Inv.* by Admiral de Witt to destroy ships' rigging.

CHAPPÉ, Claude 1794. *Inv.* semaphore signalling system connecting Paris and Lille.

CHARDONNET, Hilaire, Comte de (Fr) (1839–1924) 1884. *Pat.* process in which solution of nitro-cellulose in alcohol was pumped through fine holes of a spinneret into a bath which absorbed the solvents and left over a "rayon" thread. (Artificial silk.)

Charts (geographical and celestial) *c.* 570 B.C. *Inv.* 1400. Modern sea charts *intro.* by Prince of Portugal. 1489. *Intro.* into England by Columbus's brother Bartholomew. *c.* 1550. First fair map of England made by George Lilly. 1595. Gerard

Mercator published his atlas of maps.

Charts, Weather 1859. First synoptic weather chart issued. Admiral Fitzroy *inv.* word "synoptic." *See also* Television.

CHASSEPOT, Antoine 1868. *Inv.* breech-loading rifle used in Franco-Prussian war.

Cheque 1281. First cheque on record dated Windsor, Sept. 10, 1281.

Cheque-printing 1836. Water-colour security ("Xylographic") process *inv.* Charles C. Wright, of New York.

Chess 680 B.C. *Inv.* ascribed to Palamedes.

Chevé Musical Notation System 1818. Pierre Galin (Fr) *intro.* system founded on an earlier idea of Jean Jacques Rosseau. 1834. Aimé Paris (Fr) *intro.* "time-language" to Galin's system. 1844. Dr. Emilé Chevé (Paris's brother-in-law) compounded a complete system known as the Galin-Paris-Chevé system; which was a rival to the tonic-sol-fa system.

CHEVREUL, Michel Eugène (Fr) (1786–1889) 1823. Proved animal fats to be compounds of aliphatic acids and glycerine.

Chewing Gum A.D. 900. Mayas of Central America chewed chicle, the bases of modern chewing gum.

Chimes, Westminster 1793. Composed by Dr. William Crotch (1775–1847) for bells of St. Mary's church, Cambridge.

Chimney-sweeping Tools 1850. *Inv.* Joseph Glass.

CHLADNI, Ernst Florens Friedrich (Ger) (1756–1827)

1802. Pioneered the revival of the experimental physical approach to study of vibrations in rods, plates and membranes.

Chlorine (gas) 1774. *Disc.* by C. W. Scheele, who called it dephlogistigated marine acid air. 1784. Named "oxymuriatic acid" by Antoine Lavoisier (Fr), and "chlorine," on account of its green colour by Sir Humphrey Davy in 1810. *See also* Bleaching.

Chlorodyne (medicine) *Inv.* Dr. James Collis-Browne (1819–1884).

Chloroform 1831. *Disc.* by Soubeiran (Fr), and by Guthrie (U.S.). 1847. Composition defined by Dumas (Fr). 1847. First used as anæsthetic at St. Bartholomew's Hospital, London by Sir William Lawrence and Mr. Holmes, on the recommendation of medical student M. C. Furnell. (Results not published until later.) 1847, Nov. 15. Liverpool druggist Waldie, in co-operation with Sir James Y. Simpson of Edinburgh made personal trial of chloroform and *intro.* it to medical profession. Dr. Snow later administered Chloroform to Queen Victoria.

Chlorophyll *Disc.* and named by Pierre Pelletier (Fr) (1788–1842) and Joseph Caventon (1795–1878). 1833. Fluorescence of *disc.* Sir David Breuster. 1864. Chemical composition *disc.* 1882. T. W. Engelmann (1843–1909) *disc.* irregularity in activity of various parts of spectrum on Chlorophyll. 1960. Synthesized by R. B. Woodward, of Boston, U.S. 1906.

Composition *disc.* Richard Will-stätter (Ger).

Chloromycin *Disc.* 1947.

Chocolate 1520. Slab eating chocolate known in Spain. 1650. Sold at high prices in London coffee-houses. 1853. Popularity of chocolate in England dates from time of Gladstone's free trade budget.

Cholera 1883. Virus *disc.* by Robert Koch. *See also* Virus.

Cholesterol 1951. Synthesized from 4-methoxytoluquinone by R. B. Woodward, of Boston, U.S.

Choreography 1947. Alphabet of movement *inv.* by Rudolf Benesh.

Chorus 556 B.C. *Intro.* into Greek drama.

Christmas Cards 1862. *Intro.* by English firm of Goodall and Son.

CHRISTOFORI, Bartolommeo (It) Pioneer harpsichord maker of Padua. *Inv.* the pianoforte, 1709.

Chromatography (chemistry) 1906. *Inv.* by Twsett (Rus). 1941. Partition chromatography *inv.* Martin and Synge. 1944. Paper chromatography *intro.*

Chromescope 1869. *Inv.* by Capt. Andrew Noble (engaged by William Armstrong), and exhibited at Newcastle upon Tyne.

Chromium (element) 1797. *Disc.* by Vauquelin.

Chromo-lithography 1836. *Inv.* in England.

Chronograph, Spark 1928. *Inv.* by Loomis.

Chronometer *See* Clocks.

Chronoscope 1840. *Inv.* Sir Charles Wheatstone (1802–75).

1844. *Imp.* C. *inv.* Claude-Servain Pouillet (Fr) (1791–1868).

Chuck, Lathe 1820. Two-jawed chuck *inv.* by Henry Maudslay (1711–1832). 1820. Three-jawed chuck *inv.* Lewis Gompertz.

CIERVA, Juan de la (Sp) (1895–1936) *Inv.* autogiro *c.* 1924 *See* Aircraft.

Cigarette 1860. William and Henry Charlesworth and T. H. Dumbar, of London *pat. const.* of cigarette of leaf tobacco attached to a tube of paper or wood filled with cut or shredded pipe tobacco.

Cinematography 1872. Maybridge, of California, U.S. took 24 consecutive pictures of trotting horse. 1882. Marey, of Paris *inv.* "Marey's Pistol" which took pictures on drum-mounted glass plates (sensitized). 1893. Thomas Alva Edison (U.S.) *inv.* nickel-in-the-slot cinematograph machines used at Chicago World's Fair. 1895. First successful projector *inv.* scientific instrument maker R. W. Paul. Used in film show in Hatton Garden, London. 1899. Cinematographic *inv.* in England by W. Friese-Greene. 1895. First cine. record made by L. Lumiér (Fr) of train arriving at station. 1888. First film made by Edison. (*See* Kinetoscope.) 1906. Albert Smith proposed, then *pat.* two-colour process. 1907. Two-colour process *intro.* commercially. 1911. *Imp.* two-colour process ("Kinemacolour") *intro.* with red and green screens. 1914. John Randolph Bray *prod.* first animated cartoon—

"The Dachshund." 1916. Bray *prod.* first technical animated cartoon, and first animated colour cartoon released by Paramount. 1922. Dr. Herbert Thomas Kalmus's "Technicolor" process used to produce first successful coloured film— *Toll of the Sea.* Shown at the Rialto Theatre, New York. 1937. Frederick Waller *inv.* three-dimensional cinematography ("Viterama"), and later, with Hazard Reeves *dev.* "Cinerama" and "Cinemascope."

Circle, Quadrature of the 221 B.C. Archimedes made first approach to value of π and found the ratio as 4970 is to between 15,610 and 15,620. 1717. Abraham Sharp worked out π to 72 decimal places. 1719 Lagny worked it to 122 places (3.141592653589793238-462643383279 . . .).

Circlip (mechanics) 1820. *Inv.* by Collinge for securing lynchpins of stage-coaches.

Citric Acid 1784. *Disc.* C. W. Scheele (1742–86).

CLAIRAUT, Alexis-Claude (Fr) (1713–65) 1743. Promulgated the theorem bearing his name relating the gravity at points on the surface of a rotating ellipsoid with the compression and centrifugal force at the equator. Also put forward a theory of capillarity.

CLAPEYRON, Émile (Fr) (1799–1864) 1834. Assimilated the thermodynamic ideas of Carnot into mathematical form.

Clarionet 1690. *Inv.* Denner, of Nuremberg.

CLARK, Thomas (1801–67) *Inv.* water-softening process.

CLAUSIUS, Rudolf (Ger) (1822–88) With William Thomson (Lord Kelvin) virtually founded the science of thermodynamics.

Clavichord 17th cent. In use in France, Spain and Germany.

Clavilux (colour-organ) 1905. *Inv.* Thomas Wilfred, of New York City. 1922. First recital given in New York.

Clay, China *See* Kaolin.

CLEGG, Samuel (1814–56) *c.* 1812. *Inv.* the water-locked gas-holder, gas scrubber, gas valves, gas-meter, and other units essential in manufacture and distribution of coal-gas.

Clepshydra (water-clock) 1450 B.C. Used in Ancient Egypt with various scales for different times of the year. 158 B.C. *Intro.* into Rome by Scipio Nasica. A.D. 725. I-Hsing and Liang-Ling-Tsan (Chinese) *inv.* clepshydra with a link-work escapement.

CLERK, Dugald (1854–1932) 1881. *Inv.* two-stroke gas-engine (*Pat.* Nos. 1089–1881).

Clock 1276. Falling mercury clock *desc.* by Alfonso of Castile. *c.* 1250. First drawing of escapement principle made by Villard de Honnecourt (Fr), with model angel pointing to sun. 1250. Weight-driven mechanisms in use. *c.* 1335. Weight-driven clock made by Peter Lightfoot for abbott of Glastonbury Abbey. (Worked until 1835, when frame fitted with new gear-train.) 1335. First clock of which there is reliable knowledge set up at Milan. 1379. Clock set up in Rouen. 1386. Oldest surviving clock made. (Now at Salisbury Cathedral.) 1641 Galileo applied pendulum to clock made by his son Vicenzio. 1660. Christiaan Huygens (Hol) (1629–95) made marine clock. 1676. Rev. Edward Barlow *inv.* rack striking mechanism. 1735. John Harrison (1693–1776) made first clock (chronometer) with hand-cut gears. 1774. Thomas Mudge (1715–94) made clock (chronometer) for Board of Longitude. 1826. First clock with illuminated dials fitted St. Brides church, Fleet Street, London. *See also* Clepshydra, Escapement, Fusee, Watch.

Clock, Atomic 1946. *Inv.* Dr. Willard Frank Libby.

Clock, Electric 1840. First devised by Sir Charles Wheatstone (1802–79). 1847. *Inv.* by Alexander Bain, of Edinburgh. 1851. Electric clock for Crystal Palace *des.* Charles Shepherd. 1854. A. Hull, of Lloydsville, Ohio, took out first U.S. Pat. for electric clock. 1898. R. J. Rudd *inv.* "Synchronome" principle, with free-swinging pendulum controlling electric slave-clocks.

Cloisonné Work Devised by the Sumerians, reaching climax of perfection in ancient Egypt (Tutankhamen's collarette.)

Cloth 588 B.C. Made in Tyre. A.D. 960. Made in Flanders. 1111. *Intro.* into England. 1667. First dyeing and dressing of cloth in England.

Cloud Chamber (nuclear physics) 1911. *Inv.* C. T. R. Wilson.

Cloud Forms 1803. First classified by diarist Luke Howard, of London. (Classification substantially same as now used.)

Clutch, Friction 1786. Dog clutch *inv.* John Rennie for Albion Flour Mills, which he *des.* 1800. Cone clutch *inv.* Marc Isambard Brunel for naval block-making machinery at Portsmouth.

Coal, Underground Gasification of 1858. Suggested by William Siemens (Ger) and *imp.* by him 1883. *c.* 1890 Mendeleyer (Rus) and A. G. Bates (Eng) *imp.* process. 1912. Sir William Ramsey *inv.* new system, which was tried at Hett Gill, Tursdale Colliery, Co. Durham. 1931. U.S.S.R. subsidize experiments and build stations at Donetz and other coalfields. 1938. Russian system industrially in use. 1947. Underground gasification of coal in use in Italy. 1953. Underground gasification of coal in use at Gorgas, Alabama, U.S.

Coal-cutting Machine 1761. Horse and man-powered coal-cutting machines *inv.* Andrew Menzies (Scot). 1863. Toothed-wheel, compressed-air turbine coal-cutting machine *inv.* by Thomas Harrison. 1875. Gillot and Copley *inv.* compressed air coal-cutting machine.

Coal-tips, Hydraulically operated 1857. First *inst.* by Sir William Armstrong & Company at Roath Dock, Cardiff.

Coal Tar 1822. First recorded distillation of; at Glasgow.

Coal Washing Machine *c.* 1863. *Inv.* (with hopper conveyor) by Berard.

Cobalt (element) *Isol.* by

Kennig Brandt (Ger) (1694–1768).

Cocaine 1884. *Disc.* by Karl Kroller, of Vienna.

COCKROFT, John (*b.* 1897) 1932. With Ernest Walton *inv.* the linear accelerator to break up atoms without the use of radium.

Code, Telegraphic 1837. *Inv.* by Alfred Vail (U.S.). 1838. "Dot-dash" code *imp.* by Samuel Breeze Morse (U.S.).

Codeine 1832. *Isol.* from opium by Robiquet (Fr).

Coffee 1450. Known and used as stimulant in Abyssinia. 1643. *Intro.* in Paris. 1650. *Intro.* in Oxford. 1736. Tree transplanted from Arabia to West Indies.

Coherer (radio detector) Originally *dev.* by Prof. Thommassina (It) and *intro.* into Italian navy by Luigi Solari. 1899. Self-acting coherer *intro.* Coherer also *des.* by Edouard Branley (Fr) (1846–1940).

Coil, Induction (electrical transformer) 1838. Grafton Page (U.S.) *const.* an induction coil and *imp.* it in 1851. 1850. Heinrich Daniel Ruhmkorff (Ger) (1803–77), of Paris *const.* induction coil. 1851. Michael Faraday (1791–1867) and Joseph Henry (1797–1878) *dev.* transformer into induction coil. 1861. Induction coil used as ignition source on Étienne Lenoir's gas-engine.

Coin 1210. Silver farthing first coined in Ireland. 1257. First record of gold coin being struck in England. 1522. Silver farthing first coined in England. 1553. Half-crown first coined in England. 1663. Guinea first coined in England. 1665. Copper farthing first struck in England. (Issued 1672). 1672 (Aug. 16). Halfpenny first issued in England. 1690. Tin farthings struck in England. 1843. Half-farthing first struck in England. 1861. Half-crown discontinued in England. (Circulation resumed, 1874.)

Coining Press (mill) 1553–61. Coining press superseded hammering for minting. *Inv.* Eloye Mestrell (Fr). 1617. Coining press *intro.* England. 1790. Coining press *imp.* by Paris engraver, Chipart.

Coin-in-slot Machine Used in Ancient Greece for dispensing libation of holy water.

Coke-oven *See* Retort, Gas.

Colloids 1861. Science of colloids founded and named by Thomas Graham (1805–69).

Colon (:) 773 B.C. Colon and full-stop adopted in Greece by Thrasymachus. 16th cent. Colon and semi-colon first used in English literature.

Colour *See* Light and Optics.

Colour-blindness 1774. *Disc.* by John Dalton (1766–1844).

Combing Machine (textiles) 1832. *Inv.* Joshua Heilmann, of Mulhouse, Alsace.

Comet, Halley's 1861. *Disc.* by Edmund Halley (1656–1742), who predicted its return in 1759.

Compass, Mariner's 2634 B.C. (64th year of Chinese Emperor Ho-ang-ti.) Chinese chariot guided by piece of magnetite suspended by silk thread. 1115 and 1195. Mariner's compass also mentioned in Chinese literature. A.D. 121. Mariner's compass again mentioned in Chinese dictionary. 11th cent. First mentioned in Europe by Ara Fröde (Nor). *c.* 1100. Mariner's compass mentioned by Richard Coeur-de-Lion's foster-brother Alexander Neckham. 1242. Compass-bowl *desc.* by Bailak. 1269. First technical *desc.* of mariner's compass by Peter the Stranger. 1260 *Intro.* (?) into Europe by Marco Polo. 1302. Compass needle fixed to pivoted card by Flavio Gioja, of Amalfi, Italy. 1492. Variation of mariner's compass *disc.* Christopher Columbus. 1576. "Dip" of compass needle *estab.* Robert Norman. 1608. Box compass and hanging compass *inv.* Rev. William Barlow. 1876. Lord Kelvin (William Thomsom) (1824–1907) *pat. imp.* mariner's compass.

Compasses (drawing) A.D. 79. Bronze compasses found at Pompeii. (Compasses *inv.* Jost Bing, of Hesse, Germany.)

Compounding *See* Engine, Steam.

Compressed Air *c.* 125 B.C. Suggested as a power for catapults by Ctesibus of Alexandria. 1799. George Medhurst (1759–1829) compressed air to 210 p.s.i. and transmitted it to a motor in a mine. 1802. William Murdock (1754–1839) used compressed air from blowing machines for driving lathes, working a lift and ringing alarm bells at Soho Works, Birmingham. (The largest compressed air engine had a 12 in. cylinder and remained in use for 35 years.) 1830. Compressed air first used for tunnelling by Thomas Cochrane (1775–

1860). 1839. Compressed air used to ring signal bell between Euston Station and Camden Town, on the London and Birmingham Railway. 1861. Compressed air first used on large scale by Sommeiller (Fr). 1869. Used by Lord Cochrane for tunnelling under River Thames. *See* Drill, Compressed air.

Compton Effect (X-rays) 1923. *Disc.* Arthur Holly Compton, of Wooster, Ohio, U.S.

Computers (including **Calculating Machines**) 1641. First adding machine *inv.* Blaise Pascal (Fr) (1623–62). *c.* 1666. Sir Samuel Morland (1625–95) *inv.* adding and subtracting machine and presented it to Charles II. 1694. Gottfried W. F. Leibnitz (1646–1716) *inv.* pin-and-camwheel calculating machine. 1709. Polonius *imp.* on Leibnitz's machine. 1709. Pascal's machine *imp.* by Lepine (Fr) to carry from one column to the next. 1727. Jacob Lerpold (Ger) *inv.* calculating machine. 1774. Matthew Hahn, of Stuttgart made calculating machine in England. 1729. Hillerin de Boistissandeau *inv.* three adding machines. 1751. Jacob Pereire, of Bordeaux *inv.* adding machines for teaching deaf mutes. 1775–76. Viscount Mahon (Earl of Stanhope) built two calculating machines. 1784. J. H. Müller (Ger) *des.* calculating machine. 1820 Calculating machine re-*inv.* by Charles Xavier de Colmar (Fr). 1812. Charles Babbage (1791–1871) *inv.* his "Difference Engine," which he started building 1833 but never completed. (His son, H. P. Babbage completed a section of the machine.) C. Babbage discussed in other terms modern "programming" and "taping." 1887. "Comptometer" *inv.* by Dorr E. Felt, of Chicago, U.S. 1887. Calculating machine *inv.* by Léon Bollée (Fr). 1892. Pin-and-camwheel calculating machine re-*inv.* by W. T. Odhner (Rus). 1946. First all-electric digital computer built at Pennsylvania University. 1948. Harvard Mark II computer completed, incorporating many of Babbage's ideas. 1937. Howard Aitkin (U.S.) *des.* Harvard Mark I computer—completed, 1944. 1949. Dr. M. V. Wilkes *des.* first all-electric computer in England (18,000 valves).

Concertina 1825. *Inv.* Sir Charles Wheatstone (1802–75). *See also* Accordion.

Concordance (Bible) 1247. First compiled by 500 monks. 1737. English concordance compiled by Alexander Cruden (Scot) (1701–70).

Concrete First appearance in Roman times. Made from volcanic earth (pozzolana) found in Alban Hills, near Naples. Would set under water.

Concrete Mixer 1857. *Inv.* and used by civil engineer Cézanne (Fr) on *const.* of bridge over River Tisza at Szeged.

Concrete Reinforced 1849. First used by Joseph Monier (Fr) (1823–1906) for making tubs for orange trees. 1854. W. B. Wilkinson, of Newcastle upon Tyne embedded iron rods and second-hand mine-winding cables in concrete. 1877. Monier *pat.* reinforced concrete beams. 1892. Françoise Hennebique (Fr) *pat.* reinforced concrete beams; *intro.* into England, 1897.

CONDAMINE, Charles-Marie de la (Fr) (1701–74) 1735. With Pierre Bouguer (Fr) measured degree of earth's meridian in Peru.

Condensed Milk 1856. Borden *pat.* method of manufacturing condensed milk.

Condenser, Electric 1745. *Inv.* by Ewald Georg von Kleist, of Kamin, Pomerania; and independently by Petrus van Musschenbroek, of Leyden (1692–1761). (*See* Leyden Jar.) 1750. Benjamin Franklin (1706–90) *dev.* "Fulminating plate." 1759. Franz Alpinus *const.* electric condenser with air and glass dielectric. 1833. Bridge electroscope *inv.* by Samuel Christie. 1900. G. F. Mansbridge *dev.* paper and tinfoil electric condenser.

Condenser, Water *Inv.* by J. F. Baron von Liebig (Ger) (1803–73). 1813. Rectifying column condenser *inv.* Celier Blumenthal. Further *dev.* by Pistorius (Ger). 1834. Surface condeners *intro.* by Samuel Hall to provide fresh water for ship's engine boiler feed. (*See also* Cooling.)

Conic Sections 330 B.C. Aristotle (384–322 B.C.) wrote first treatise on conic sections.

Connecting-rod *c.* 1430. Applied to crank as substitute for human arm by a German millwright. (*See also* Crank.)

CONTI, Abbe Anthony Schenella (It) (1677–1749)

Prop. principle of aneroid barometer.

Contour Lines (map) 1791. *Inv.* and first used by French surveyor Jean Louis Dupont-Triel.

Contra-bassoon *Inv.* Wilhelm Heckel (Ger).

Controls, Flexible (engineering) 1896. Twisted steel wire cable running in spiral spring "tube" *inv.* E. M. Bowden.

Conveyor 1794. William Staden *inv.* conveyor claiming to "work upon a plane, an ascent, or perpendicular." 1795. Oliver Evans (U.S.) (1755–1819) *desc.* canvas belt conveyor in his book *The Miller's Guide.* 1903. Canvas sling conveyor *const.* for use at Mobile, Ala., U.S. 1910. Drew and Clydesdale built first conveyor belt in Britain for bananas; *des.* by Percy Donald.

Conveyor, Coal-face 1902. First satisfactory coal-face conveyor *inv.* W. C. Blackett; with steel trough and scraper-chain.

COOKSWORTHY, William 1760. *Disc.* kaolin in Cornwall and made first hard porcelain in England.

COOPER, E. A. 1878. *Inv.* Teleautograph or writing telegraph.

COOPER-HEWITT, Peter (U.S.) 1902. *Inv.* Mercury arc rectifier and mercury vapour lamp.

COPERNICUS, Nicholaus (1473–1543) 1543. Explained his theory of the universe with the sun in the central position, instead of the earth, as in the Ptolemaic system.

Copper *c.* 4200 B.C. First used. 1561. First *disc.* (modern) in Cornwall. 1869. Electrolytic refining commenced at Pembray, South Wales, using Elkington's process. 1878. Holloway's refining process first used. *See also* Electrotyping.

Copper Arsenate (Scheele's Green) 1778 *Disc.* by Carl Wilhelm Scheele (Ger) (1742–86).

Copper-plate Printing *See* Printing.

Copper-plating of ships' bottoms *c.* 1860. *Intro.* during American War of Independence.

Copying Machine 1778. Copying machine for writing *inv.* by James Watt (1736–1819). 1845. Panel copying machine for wood and stone used on British Houses of Parliament. Also *pats.* by Bruncl and Wedgwood.

Cordite *Inv.* F. A. Abel and J. Dewar.

Coriolis Effect (geophysics) 1860. First suggested and analysed by Gustave-Gaspard Coriolis (Fr) (1792–1843).

CORIOLIS, Gustave-Gaspard de (Fr) Mathematician. *See* previous entry.

CORLISS, George Henry (U.S.) (1817–88) 1849. *Inv.* steam-engine valve-gear bearing his name.

CORNELISZ, Cornelis (of Uitgeest) 1593. *Inv.* wind-driven saw-mill and seed-crushing, edge-runner mill.

CORRENS, Karl Erich (Ger) (1864–1933) *Imp.* the electrical accumulator *inv.* by Planté in 1859, and with Faure, *dev.* the storage battery used in modern motor-car.

Corrosion Fatigue (metallurgy) 1917. Phenomenon first

identified (in brass) by Haigh.

Corrugated Iron 1832. *Inv.* John Walker, of Rotherhithe, London and used for roofing. (Cost £5 10s. per 100 sq. ft.)

CORT, Henry (1740–1800) 1783. With Peter Onions *inv.* the puddling process for conversion of pig-iron into wrought-iron.

Cortisone 1935. *Isol.* by Dr. Edward C. Kendall (U.S.). 1948 (Sept.). First used on human patient (a woman). 1951. Synthesized by R. B. Woodward, of Boston, U.S.

Cosmic Rays and Radiation 1910. Gockel (Swit.) sent up electroscope to 12,000 ft. to check cosmic rays. 1911. Kolhorster and Hess check cosmic rays up to 15,000 ft. 1914. Kolhorster checks cosmic rays to 27,000 ft. *c.* 1920. R. A. Millikan and Bowen check cosmic rays to 45,000 ft. with self-registering machines. 1927. *Disc.* that cosmic radiation is greater at high latitudes than at equator.

Cotton Gin (textiles) 1793. *Inv.* Eli Whitney (U.S.) (1765–1825). 1796. *Imp.* by Holmes.

Cotton-picking Machine 1889. Angus Campbell (Scot) *inv.* spindle-type cotton-picking machine. 1925. Hiram M. Berry (U.S.) *inv.* barbed-spindle cotton-picking machine. 1933. John and Mack Rust (U.S.) tested their first spindle-type cotton-picking machine.

COULOMB, Charles Augustin de (Fr) (1736–1806) 1785. Formulated the law of forces between electrical charges which is named after him.

Coumarin (chemistry) 1868. First produced by Sir William Henry Perkin (1838–1907).

Counterbalance Weight c. 1480. *Ref.* to ball of metal on one crank to counterbalance the other, found in German literature. 1845. First applied to locomotive steam-engine by William Fernihough.

COURTOIS, Bernard (Fr) (1777–1838) 1811. *Disc.* iodine.

Cowl, Chimney 1714. Revolving cowl *inv.* in France. 1850. Downdraught, non-smoking cowl *inv.* by William Pilbeam.

COWPER, Edward (*c.* 1857) 1857. With John M'Naught *intro.* modern rotative form of steam-engine.

CRAFTS, James Mason (1839–1917) 1877. With Charles Friedel paved the way for large-scale chemical synthesis by devising methods which made possible the substitution or introduction of certain groupings of atoms into organic compounds.

Crane (mechanics) *c.* 1550 B.C. Lifting crane in use in Egypt. *c.* 50–26 B.C. Sheerlegs type crane in use in Rome. 1520 (A.D.). Cranes *illus.* by Georg Bauer (Georgius Agricola) (1494–1555). 1714. Balanced, mast-mounted ship's cargo derrick crane *inv.* Père Ressin (Fr). 1785. Jean Tremel (Fr) *imp.* ship-loading crane. 1790. Ambrose Poux-Landry (Fr) *imp.* on Tremel's *inv.* 1792. Pierre Desvallons (Fr) *inv.* two types of ship-loading cranes, one of which enabled one man to handle 2½ tons. 1790. Joseph Bramah (1748–1814) *des.* wall-type warehouse crane. 1790. Joseph Bramah (1748-1814) *des.*

wall type warehouse crane. 1790. Weighing crane *inv.* by Andrews. 1795. Crane with travelling carriage *des.* by Gottleib. 1846. First hydraulically operated crane erected at Newcastle upon Tyne. 1850. 50-ton crane erected at Glasgow. 1858. Steam-operated overhead travelling crane in use in Whitworth's London works. (In use as lately as 1930.)

Crane, Weighing 1800. *Inv.* by Andrews.

Crank (mechanics) 37–41 (A.D.) Cranks and flywheels *disc.* on Roman Emperor's ceremonial barge recovered from Lake Nemi, Italy. 816–834. The Utrecht Psalter (at Rheims) depicts earliest European crank and earliest rotary grindstone. 942. Abbot Odo de Cluny (*d.* 942) *desc.* crank-operated organistrum or hurdy-gurdy. 1206. Al Jazari (Arab) mentions application of crank. 1335. Italian physician Guido da Vigevano *desc.* combination of two Luttrell Psalter-type cranks to form one crank in centre of axle. 1340. Double-cranked grindstone mentioned in Luttrell Psalter. 14th cent. Large two-man, cranked crossbow in use. 1405. Konrad Keyser pictures five cranked devices for cross-bows. 1420. Practical compound crank *intro.* in form of carpenter's brace. *c.* 1430. MS. shows mill operated with crank, connecting-rod and treadles, with a flywheel. 1556. Georg Bauer (Georgius Agricola) (1494–1555) mentions crank as means of working windlasses and pumps. 1780. Crank re-*inv.* by Birmingham button-maker James Pickard. (*Pat.* expired, 1794.) 1769. James Watt (1736–1819) mentions his use of crank. 1865. Ernest Michaux, of Paris, first applied cranks to cycle propulsion. (Karl Kech, of Munich, is said to have made this application independently the same year.)

CRAWFORD, Adair (1748–95) With chemists and physicists Gadolin, Kraft, Richmann, Wilcke and Robinson continued research work in calorimetry along the lines laid down by pioneer Joseph Black (1728–99).

Crayon 1748. First made by L'Oriot (Fr).

Creatine (chemistry) 1835. *Disc.* by E. Chevreul (Fr).

Creosote 1833. *Disc.* by Reichenbach (Ger).

Creosoting (wood preservation) 1838. Process *inv.* by Bethell. *See also* Kyanizing.

Crêpe (textiles) A.D. 680. First made at Bologna, Italy.

CROMPTON, R. E. B. (1848–1940) *c.* 1870. *Des.* steam road vehicle used in India.

CROMPTON, Samuel (1733–1827) 1774. *Inv.* spinning "mule."

CROOKES, Sir William (1832–1919) 1873. *Inv.* the radiometer.

Crops, Rotation of A.D. 763. Three-field system of agriculture first mentioned.

CROSS, Charles Frederick (1855–1935) *c.* 1915. *Inv.* petroleum cracking process. 1882. *Inv.* (with Bevan) a good substitute for cotton by making a regenerated cellulose yarn from a solution of cellulose-xanthate (formed from cellulose, lye, and carbon disulphide, to be spun into viscose-rayon fibre.

Cross-bow 200 B.C. The "scorpion," a *dev.* of the cross-bow widely in use. *See also* crank.

Cryogenics Science pioneered by Z. F. von Wroblewaky in 1883. *See* Gases, Liquefication of; Refrigeration; Refrigerators.

Cryophorus (latent heat) *Inv.* Dr. William Hyde Woolaston (1766–1828).

Crystallography 1669. Nicola Steno first noted regularity and constancy of facial angles of crystals. (Constant form in metals had been *disc.* by Andrew Caesalpinus (It) (1519–1603).) 1771. Science of crystallography founded by Romé de Lisle, who, in 1783 made instruments to measure angles of crystals. 1801. Modern school of crystallography founded by Réné-Just Haüy (Fr).

Crystal Field Theory 1930. *Dev.* by Berthé and Van Vleck.

Crystal Rectifiers *See* Transistors.

CTESIBOS Alexandrian philosopher (*c.* 300 B.C.) *inv.* water-blown hydraulos, or musical organ.

Cudbear (archil, orchil; dye) Type of dye obtained from certain litchens *pat.* by Dr. Cuthbert Gordon, who connected it with his name, *c.* 1840.

CUGNOT, Nicholas Joseph (Fr) (1725–1804) 1769. *Des.* and *const.* first steam vehicle to run under its own power on a common road.

Cuirass (armour) 1216. First worn by English cavalry.

Cultivator, Rotary 1853. *Pat.* by Chandon Wren Haskyns.

Cuneiform Writing 1800. Grotefend fortuitously translated the name "Darius" and "Xerxes" = "Kschersche," his son. 1844. Lassen completed a cuneiform alphabet. (Bernouff (1775–1844) and Sir Henry Rawlinson (1810–95) also deciphered cuneiform writing.)

Cupellation 2500 B.C. Process used to refine gold. Also used by the Romans.

CURIE, Pierre (Fr) (1859–1906) 1898–1902. *Disc.* and studied many radioactive elements. (*See also* Becquerel.) 1898. *Isol.* polonium, and in 1912, radium as a pure salt.

CURIE, Marie (Fr) (1867–1934). 1898. Jointly with her husband, Pierre, *disc.* radium.

Curing (salting) 1260. William Beukelszoon, of Biervleit, Holland, first cured herrings by gutting and salting. 1359. First applied in England. 1397. Used for herrings.

Curium (element) 1945. *Disc.* by Glen T. Seaborg (U.S.), R. A. James (U.S.) and A. G. Liorso (U.S.).

Curtal (musical instrument) *See* Bassoon.

CURTIS, Charles G. (U.S.) (1860–1953) 1896. *Inv.* velocity compounding in steam turbine design.

Cutting-out Machine (textiles) 1853. Frederick Osbourn *inv.* first cutting-out machine for clothes.

CUVIER, Georges (Fr) (1769–1832) Virtually founded the science of palæontology and enunciated the principle of correlation.

Cyanamide (chemical) 1851. *Disc.* by Cloez and Stanislao Cannizaro (It) (1826–1910).

Cyanogen (chemical) 1815. First obtained in free state by Joseph-Louis Gay-Lussac (Fr) (1778–1850).

Cyanogen sulphide (chemical) 1829. *Disc.* Baron von Liebig (Ger) (1803–73).

Cycle (bicycle) 1790. First hobby-horse *inv.* Chevalier de Sivrae (Fr); according to "Le Cyclisme Théoretique et Pratique." 1816. Baron Karl Drais von Sauerbronn of Karlshrue *inv.* hobby-horse, or "Draisene" ("Célérifère"). (*See also* Niépce.) 1819. London coach-maker George Birch *const.* a hand-lever-propelled cycle on which he "rowed" 67 miles in one day. 1835–38. Kirkpatrick Macmillan, of Pierpoint, Dumfries-shire, built and rode a cycle of his own *des.* 1868. Word

"bicycle," (spelt "bysicle") used first in the London *Times.* 1885. Safety cycle *dev.* James Starley. *See also* Crank, Michaux, Starley.

Cycloidal Curve. 1451. First mathematical analysis of cycloidal curve published.

Cyclone (meteorology) 1848. Word coined by Piddington in his *Sailors' Horn-book.* 1861. Term anti-cyclone first used by Sir Francis Galton in his *Meteorographica.*

Cyclotron (magnetic resonance accelerator) 1936. *Inv.* Ernest O. Lawrence (U.S.) (1901–).

Cymbal 1580 B.C. *Inv.* ascribed to Cybele.

Cypher Writing 400 B.C. Used by the Spartans.

Cystoscope (medical) 1877. *Inv.* by Max Nitz (1848–1906), of Berlin.

Cytology 1882. Initiated by Walther Flemming (1843–1915), of Prague.

D

DAGUERRE, Louis Jacques Mandé (Fr) (1789–1851) 1839. Perfected production of silver photographic image on a copper plate.

Daguerretype (photography) *See* previous entry.

Dahlia (botany) 1784. *Disc.*

Dr. Dahl (Swed) in Mexico. *Intro.* England 1814.

DAIMLER, Gottleib (Ger) (1834–1900) 1885. Made his first motor-cycle of ½ h.p. 1885. Made second single-cylinder internal combustion engine. 1886. Made first four-wheeled motor vehicle—$1\frac{1}{2}$ h.p. single-cylinder engine. 1889. *Pat*. Vee-twin engine, and licences Pan-hard-Levassor (Fr) to make and use it. 1891. Builds first motor lorry. *See also* Motor-car and Engine, internal combustion.

DALTON, John (1776—1844) 1802–03. *Prop*. atomical theory.

Damask (linen) 1571. First made in England.

Dam (hydraulics) 1300 B.C. $1\frac{1}{4}$ mile long stone dam built in the Orontes Valley, Syria.

DANIELL, John Frederick (1790–1845) 1836. *Inv*. electric battery bearing his name.

DARBY, Abraham Senior (1677–1717) 1709. First made "coak" from "cole" and in 1717, first used it for iron smelting. (*q.v.*)

DARWIN, Charles (1809–1882) 1842. *Prop*. theory of evolution.

DAVENPORT, Thomas (Scot) 1839. *Inv*. small electric motor and drove workshop tools with it.

DAVEY, Sir Humphry (1778–1829) 1808. *Disc*. elements strontium, calcium, magnesium and barium.

DAWES, William Rutter (1799–1868) 1850. With Lassell and Bond *disc*. "crêpe" ring of planet Saturn.

"D.D.T." (insecticide) 1874. First *prep*. by Othman Zeidler (Aus) with no idea of its insecticidal properties. 1939. Paul Muller (Swit) *disc*. its insect-killing power.

Deaf Aids 1705. Horn type *inv*. by Du Guet (Fr). Carmelite monk Père Sebastien (Jean Truchet) (1657–1729) also *inv*. a trumpet type deaf aid.

Decimal System *c*. 1450 Puerbach *inv*. a decimal system. 1585. Simon Stevens (Stevinius), of Bruges, *inv*. a decimal system. Wrote book : *La Decimal*. 1795. Decimal system adopted in France.

Declination, Magnetic 1702. Edmund Halley *pub*. first world chart of magnetic declination.

DE FOREST, Dr. Lee (U.S.) *Inv*. three-electrode radio valve.

DEIMAN, Joan Rudolf (Hol) (1743–1808) Analytical chemist, who, in 1789 (with Adriaan Paetz van Troostwijk) *disc*. that water was decomposed when a battery was discharged through it.

DE LA CAILLE, Nicholas-Louis (Fr) (1713–62) 1763. *pub*. a catalogue of 2,000 stars from the 10,000 he *obs*. during his stay at the Cape of Good Hope.

DE LA RUE, Warren (1815–89). First to demonstrate that electric current passing through a coil of (platinum) wire, caused it to glow to give light.

DELUC, Jean-André (Fr) (1727–1817) 1751. *Inv*. a type of hygroscope.

Dentures 1585. First mentioned in England. 1609. In more general use. (1399. London barbers awarded 6d. per day by Henry IV to draw teeth of the poor at no charge.)

DE RIVAZ (Fr) of Sion, Valais 1807 (Jan. 30). *Pat*. "Machines dont le principle moteur est l'explosion de gaz et autre substances."

DEROSNE, Charles (1780–1846) Perfected modern rectifying column for alcohol production.

Derrick, Drilling 1830. First *intro.* 1850. Steam-drive *intro.* 1857. Derrick-drilling used in Hanover. 1860. Diamond drill-head *intro.* 1895. Galician A. Raky *intro.* first fast drilling-rig.

DESAGULIERS, J. Théophile (Fr) (1685–1744) Experimental physicist, particularly in static electricity.

DESCARTES, René du (Fr) (1596–1650) of Tours *Prop.* Cartesianism system of philosophy. Founder of modern philosophy.

DESLANDRES, Henri Alexandre (Fr) (1853–1948) 1890. *Inv.* spectro-heliograph.

DESORMES (1777–1842) With Lavoisier, Laplace, Clemént, Delaroche, Bérard and Renault made early accurate measurements of the specific heat of gases at constant pressure and volume.

Detector, Radio 1906. Carburundum/steel detector *inv.* Gen. Dunwoody (U.S.). This was the first crystal radio detector. Jigger, or oscillation transformers *inv.* Marconi, Slaby and Sir Oliver Lodge. Plumbago/galena detector *inv.* in Ger, also galena/tellurium and silicon/steel, by Otto von Bronk and Pickard, respectively. Walter and Ewing also *inv.* magnetic detectors. Self-restoring detector with greasy steel wheel *inv.* Lodge-Muirhead. Electrolytic detectors *inv.* Ferrie, Fessenden and Schloemilch. Thermo-electric detectors *inv.* Fessenden, Diddell, Fleming and Austin. Hessite (silver telluride) and anastase

(native titanium oxide) *inv.* Prof. Pierce.

Deuterium (heavy hydrogen) 1932 (Jan. 1). *Disc.* H. C. Urey. 1932 (later). Deuterium of great purity separated from hydrogen by multiple diffusion by G. Hertz.

Dew 1814. First theory of the formation of dew *prop.* by Dr. Charles Wells.

DEWAR, Sir James (1842–1923) *Inv.* of the vacuum flask, and, with Sir Frederick Abel, cordite.

Diabetes *c.* 2nd cent. A.D. *Disc.* and named by Aretaeus of Cappadocia, who lived at Alexandria.

Diagnosis, Medical Evolved by Alpinus (1553–1617).

Dialysis 1862. First studied by Thomas Graham.

Diamagnetism *Disc.* Michael Faraday (1791–1867), and further studied by Alexandre Edmond Becquerel (Fr) (1820–91).

Diamond 1491 B.C. Earliest mention of diamond, Exodus xxvii, 18. A.D. 1456. Diamond polishing art *disc.* Louis Berquen, or Berghen, of Bruges. 16th cent. Engraving on diamonds first attempted. 1694. Averami and Targioni (It) fused a diamond at the focus of a burning-glass. 17th cent. (end of). First brilliants cut by Peruggi, of Venice. 1727. Diamonds *disc.* in Brazil. *c.* 1771. Antoine-Laurent Lavoisier and Macquer (Fr) proved diamond to be carbon by burning one in oxygen and collecting the CO_2 thus formed. 1867 (Mar.). Diamonds *disc.* at the Cape of Good Hope.

Diamond, Artificial 1880. Diamond synthesized Hannay. 1897. First made by Henri Moissan (Fr) (1852–1907). 1908. Synthetic diamonds made by Vicompte E. de Bois-menu, of Paris.

Diatoms 1773. First *desc.* by Otto Friedrik Müller (Dan) (1730–84). 49 types *desc.* by Carl Adolf Agardh (Swed) (1785–1859).

Diazo Compounds 1858-64. *Disc.* Griers (Ger) working at English brewery.

Dichlorethylene (chemistry) 1840. Henry Victor Regnault (Fr) (1810–78) announced a strange new fluid he had *disc.*, which was dichlorethylene. 1922. Regnault's fluid identified as Dichlorethylene. 1940. Dichlorethylene first used as commercial plastic and renamed "Vinylidene" chloride.

Dichröiscope 1860. *Inv.* Prof. Dove, of Berlin.

Dictionary 1100 B.C. Earliest by Pa-out-tse (China). *c.* 60 B.C. Earliest Latin dictionary compiled by Marcus Terentius Varro (A.R.) (117–27 B.C.). 1755 Dr. Johnson compiled his Great English dictionary. 1791. Walker's English Dictionary published.

DIRAC, Adrien Maurice (Fr) 1925. *Disc.* significance of "Poisson's Brackets" (maths.).

Dirigible (aircraft) 1784. Brisson (Fr) advocated cylindrical balloon with conical ends. 1784. Duc de Chartres (Fr) *const.* Brisson-type balloon and ascended at St.Cloud with Roberts Brothers and Colin-Hulin. 1785. Prof. Kramp, of Strasburg (Fr) *des.* egg-shaped dirig-

ible with balloonets. 1850. First model dirigible to fly under power *const.* by Pierre Jullien (Fr) (23 ft. long). 1852. Jullien *const.* full-size dirigible "Le Precurseur," 164 ft. long. 1852. Henri Giffard (Fr) (1825–82) *const.* 143 ft. long dirigible with rudder and 3 h.p. steam-engine; and made 17 mile flight from Hippodrome, Paris to Trappes. 1872. Paul Haenlin, of Vienna *const.* dirigible powered by a Lenoir gas-engine and propeller. 1872. Depuy de Lôme (Fr) *const.* egg-shaped dirigible for French Government. 1900. Count Zeppelin (Ger) built his first dirigible and flew it over Lake Constance. 1921. British dirigible "R38" broke up over River Humber. 1924. U.S. dirigible "Shenandoah" broke up over Ohio. 1930. British dirigible "R-100" wrecked and burnt out at Beauvais, France. 1933 U.S. dirigible "Akron" lost off New Jersey coast. 1936. German dirigible "Hindenburg" burnt out on landing at Lakehurst, U.S. (*See also* Aircraft, Balloon.)

Displacement, Law of (optics) 1893. *Prop.* by Wilhelm Wien (Ger).

Dissection (surgical) *c.* 1300. Earliest known *illus.* of operation.

Distillation 350 B.C. Aristotle records experiments for removing salt from sea-water by distillation. 49 B.C. Julius Caesar used distillation to supply legions with drinking water. A.D. 1. Distillation first used in Spain to obtain mercury from mercury sulphide. (*See also* Cupellation and Amalgams.)

1150. Water distillation *intro.* into Europe by the Moors. 1801. Edouard Adam (Fr) *inv.* apparatus for distilling brandy from wine. 1808 Blumenthal *inv.* continuous still, simplified by Langier (Fr). 1813. Blumenthal Cellier *des.* and *const.* first rectifying column for separating alcohol from wine. (*Column* perfected by Charles Derosne and Savalle (Fr).) 1832. Analysers and rectifiers *inv.* Æneaus Diblin and Coffley. *See also* Condenser, Water; and Radiator Motor car.

Distribution, Law of *Prop.* (*disc.*) by Clerk Maxwell (Scot) (1831–79).

Dividers (Draughtsman's) 79 (A.D.). Bronze dividers in use in Pompeii.

Diving Bell 325 B.C. Mentioned by Aristotle (384–322 B.C.). 320 B.C. Used in Phoenicia. 1509. Mentioned by John Taisnier of Hainault as being used at Toledo (Cadiz) before Emperor Charles V of Spain. 1716. Edmund Halley (1656–1742) first used diving bell for deep-sea diving. 1779. John Smeaton (1724–92) first used diving bell for engineering works at Ramsgate Harbour (1788).

Diving Dress 1668. Suggested by Giovanni Alphonse Bonelli (1608–79) (It). Goat-skin and 2-ft. diameter copper headpiece used, with pipes to stop condensation. 1772. Idea of diving dress (helmet) suggested to Périer (Fr) by a barber. Périer dived successfully with it into 50 ft. of water. 1831. Diving dress *inv.* by Dean for salvaging wreck of *Royal*

George. 1836. Diving dress *pat.* by Siebe. 1838. Siebe's diving dress first used in England. 1880. Fleuss greatly *imp.* diving dress and brought it into practical use in Severn Tunnel and Killingsworth Colliery disasters.

D.N.A. (Deoxyribonucleic acid) 1869. Friedrich Müller (Swit) *disc.* nucleic acid, which was later renamed D.N.A. 1953. Re-*disc.* by Dr. Maurice Wilkins, of King's College Hospital, London. (*See also* R.N.A.)

DOLLOND, John (1706–61) 1755. *Inv.* achromatic lens.

DÖPPLER, Christian Johann (Aus) (1803–53) 1842. *Disc.* that sound waves increase or decrease in pitch as their source moves towards or away from observers—known as the "Döppler Effect."

Döppler Effect 1845. Effect verified by Dutch meteorologist Christopher Buys Ballot. (*See* previous entry.)

"Dracone" *See* Barge.

DRAPER, John William (1811–82) 1847. *Disc.* infra-red rays.

DREBBEL, Cornelius (Hol) (1572–1634) 1598. *Pat.* a submarine and a thermostatically controlled egg incubator.

Dredge, Naturalist's 1773. *Inv.* Otto Frederick Müller (Dan) (1730–84).

Dredger 1561. Bucket dredger *pat.* by Vehturino, of Venice. 1561. Bucket dredger *const.* by Pieter Breughel for use on Rupel-Scheldt canal, for Brussels Municipality. 1640–1700. "Amsterdam" power dredger operated by horses. 1796. Boulton and Watt 4 h.p.

ladle dredger in use at Sunderland. 1804. Steam dredger with bucket chain *dev.* by Oliver Evans (U.S.). 1805. George Rennie (1791–1866) *des.* dredgers used on River Clyde (1824), and River Ribble (1838). 1863. Trough dredger *inv.* by Lavelley (Fr) for Suez Canal. 1871. Centrifugal pump dredger in use in U.S.

Drift, Continental (geology) 1915. Theory *prop.* by Wegener (Ger).

Drift Net 1416. *Intro.* by Dutch (360 ft. long).

Drill 2750 B.C. (Neolithic period). Fire drill in use. 2500 B.C. Bow-drill in use in Egypt. 1450. Triple bow-drill used by bead-makers of Ancient Egypt. *c.* 70 (A.D.). Ancient Roman drill-bits found in Pompeii. James Nasmyth (1808–90) *inv.* portable hand-drill. *c.* 1844 Sir Joseph Whitworth *des.* pillar sensitive drilling-machine. 1850. Pillar drill in general use. 1851. Radial drill shown at Great Exhibition, London. 1750. Crude pillar-drill in use. 1850. Flat, arrow-headed drills in use in U.S. 1860. Morse type twist drills *inv.* 1860. Hydraulic-feed drilling-machine *inv.* John Cochrane. 1887. F. J. Rowan *inv.* electric drill for shipbuilding. 1892. J. H. Griffith *inv.* double-channel archimedean drill.

Drill, Rock 1813. First *inv.* Richard Trevithick. 1857. Compressed-air rock-drill *inv.* French engineer Sommeiler. 1871. Ingersoll drill *inv.* 1880. Hydraulic rock-drill *inv.* Brandt.

Drill, Slotting 1817. *Inv.* Roberts.

Drip, Saline (medicine) *Inv.* John Murphy (1857–1916), of Appleton, Wisc., U.S.

Drugs *See* individual drugs by name: e.g. Adrenaline, Salvarsan, etc.

Drum 713. *Intro.* into Spain by the Moors.

Drum Tuning 1812. Thumbscrew method of drum tuning *inv.* by Gerard Cramer, of Munich.

Drydock 1641. "Ship's Camels" appeared. 1688. Bakker (Hol), of Amsterdam, *des.* more efficient drydock.

Dualistic Theory (chemistry) 1811. *Prop.* by Jöns Jacob Berzelius (Swed) (1779–1848).

Duct, Thoraic 1563. *Disc. in* horse by Eustacius. 1664. In human being by Ol Rudbec (Swed) (also by Thomas Bartholine (Dan) and Dr. Joliffe (Eng)).

DUFAY, Charles-Francois de Cisternay (Fr) (1698-1739) *c.* 1703. With Jean-Jacques d'Ortous, and René-Antoine Ferchault and Réaumur pioneered the study of the phenomenon of electroluminescence.

DULONG, Pierre Louis (Fr) (1785–1838) 1819. With A. T. Petit *pub.* a paper interlinking the atomic theory with the theory of heat.

DUMAS, Jean Baptiste André (Fr) (1800–84) 1830. Perfected methods of incinerating small amounts of organic compounds and absorbing the combustion gases in chemicals to determine the percentage of carbon, hydrogen, nitrogen,

etc., of the original substance.
Duplicator (transfer printing)
1780. Glutinous ink type *inv.* by
James Watt (1736–1819).
Duralumin (alloy) 1906. Accidentially *disc.* and *pat.* by Alfred
Wilm (Ger) and originally produced at Duren, at Durence
Metal Works; hence its name.
DÜRER, Albrecht (Ger)
(1471–1523) Pioneer of translation of Latin science books
into national languages.
Dusting Machines *See* Bolting
machines.
Dyeing 3000 B.C. Indigo dyed
garments *disc.* in Thebes tombs.
2000 B.C. Art of mordanting
disc. 1491 B.C. Dyeing mentioned in the Bible (Exodus
xxxv. 23). 1429. First book on
dyeing published in Venice.
1557. Dyeing with indigo prohibited by decree of German
Diet. *c.* 1560. Use of indigo
prohibited by Queen Elizabeth.
1608. Dyeing *intro.* into England from Holland. 1685. Fugitive dyers settle in England.
1628. Two Exeter dyers flogged
for teaching the art in the north
of England. 1765. Dyeing
Turkey red *intro.* into Scotland
by Papillon (Fr). 1848. Mauve
dyeing (from lichens) *intro.* by
Marnas (Fr).
Dyes, Synthetic 1856. Magenta (first aniline dye) *disc.* Sir
William Henry Perkins (1838–
1907). 1773. First synthetic dye
(yellow) for cotton *inv.* Dr. R.
Williams. 1876. Methylene
blue *disc.* Heinrich Caro (Ger).
1878. Methylene green *disc.*
Emil Fischer (Ger). 1880. Synthetic dye (indigo) *disc.* Adolf
von Bayer (Ger). 1884. Synthetic dye (Congo red) *disc.*

Böttger (Ger). 1887. Synthetic
dye (primuline) *disc.* by
Greene.
Dynamite 1867. *Inv.* by Alfred
Nobel (Swed) (1833–96) by
combining nitro-glycerine with
kieselguhr or infusorial earth.
1868 (July 14). First tried for
blasting at Greystone Quarries,
Merstham, Surrey.
Dynamo 1831. First dynamo
inv. by Hippolite Pixii (Fr),
with rotating magnet and two
fixed coils; producing alternating current. 1876. C. F.
Brush (U.S.) built his first
dynamo. (*See also* Motor, Electric.)
Dynamometer 1821. Absorption type (friction-brake) dynamometer *const.* by Piobert and
Hardy. Known as the prony
dynamometer *inv.* by Gaspard
de Prony (Fr). 1833. Prony
dynamometer first commercially used for testing output of a
small Fourneyron turbine (*q.v.*)
in France. 1860. Dynamometer
inv. by Col. Morin (Fr). *c.* 1860.
Dynamometer *inv.* by Gen.
Poncelet (Fr). (*See* Turbine,
Water.) 1865. Hirn used centrifugal pump (*q.v.*) as dynamometer. 1877. William Froude
inv. hydraulic friction-brake
dynamometer.
Dynasphere (vehicle) 1937.
Mono-car with 10 ft. wheel *inv.*
Dr. J. A. Purves. It could travel
at 30 m.p.h.
Dysprosium (element) 1886.
Existence proved by Lecoq de
Boisbaudran (Fr). 1906. Obtained pure by G. Urbain by
fractional crystallization process. (Ion exchange process
now used.)

E

Ear 1771. Auditory organs of fishes *disc.* by Pieter Camper (Hol) (1772–89). *See also* Eustacian tube.

Earth, Electrical conductivity of the Fact *disc.* Carl Augustus Steinheil (Ger) (1801–70).

Earth, Density of the 1774. Specific Gravity determined by Nevil Maskelyne (1732–1811). (Later confirmed by Cavendish, Baily, Sir Heney James, Sir Edward Sabine and others.)

EASTMAN, George (U.S.) (1854–1932) 1885. *Inv.* machine for manufacturing photographic paper in long rolls.

Eau de Cologne *c.* 1850. *Inv.* by Jean Marie Farina, of Cologne.

Eccentric *c.* 1810. *Pat.* by William Murdock (1772–1847).

Echo Sounder *Intro. c.* 1925.

Ecliptic, Obliquity of the *Disc.* Eratosthenes (276–196 B.C.) by use of armillary spheres.

Ecology 1886. Word *intro.* by Haekel.

Economiser, Steam-engine 1843. Feed-water heating *inv.* by Edward Green.

EDGEWORTH, Richard Lovell (1744–1817) With J. L. Macadam and John Metcalf pioneered new methods of road *const.*

EDISON, Thomas Alva (U.S.) (1847–1931) *Inv.* phonograph, carbon microphone, mimeograph, electric pen, alkaline accumulator and many other *invs.* (all of which *q.v.*).

Edison Effect (electrical) 1883. *Disc.* Thomas Alva Edison. Edison effect explained by Sir Alexander Fleming (1881–1955), who therefrom *dev.* the thermionic valve (*q.v.*).

EINSTEIN, Albert (Swit) (1879–1955) 1905. First *prop.* the theory of relativity.

Einsteinium (element) 1952–53. *Disc.* by Glen T. Seaborg and A. G. Liorso (U.S.).

Ejector Pump (water) 1852. *Inv.* James Thomson.

Elasticity 1798. Dr. Thomas Young (1773–1829) gave name to modulus of elasticity. 1829. Simeon-Denis Poisson (Fr) (1781–1840) *disc.* ratio bearing his name. *See also* Tensile-test machine.

Electricity *c.* 1600. Made a science by Dr. William Gilbert (1544–1603). *See* individual applications.

Electricity, Velocity of Measured by Sir Charles Wheatstone (1802–75) and determined to be 288,000 miles per second.

Electricity, Animal 1756. Marc. Leopoldo A. Caldini (It) *obs.* frog's leg contract under influence of electricity. 1760. Sensation of taste produced by silver and lead touching in the mouth noted by Johann Georg

Sulzer (Swit). 1791. Luigi Aloisio Galvani (It) (1727–98) noted effect of electricity on muscular motion.

Electricity, Atmospheric 1747. Dr. Benjamin Franklin (U.S.) (1706–90) *pub.* his *discs.* on atmospheric electricity including his experiences with kite. 1752. First *obs.* by Louis-Guillaime Le Monnier (Fr) (1717–99).

Electricity, Static Machinery for production of pioneered by John Canton.

Electric Generating Station 1882 (Jan. 12). First public supply electric generating station in world *des.* by T. A. Edison at 57, Holborn, London.

Electric Meter First *inv.* by Lord Kelvin 1872. Mechanical electric meter *inv.* by Samuel Gardiner.

Electro-cardiagraph *Intro.* after method of recording electrical impulses of heart had been *disc.* in 1879 by Dr. Augustus Waller (1816–80), of Kensington, London. 1856. Electrical currents of heart *disc.* and demonstrated by Koelliker and Müller.

Electro-deposition of metals *See* Electro-plating.

Electro-luminescence *c.*1703. Study of pioneered by C. F. Dufay, J. J. d'Orlöns, R. A. Ferchault, and R. A. F. Réaumur. 1936. Demonstrated by Destrain.

Electrolysis 1789. Johan Rudolf Deiman and Adriaan Paetz (Hol) *disc.* water decomposed by passage of electric current. 1800. Carlisle and Nicholson decomposed water by electricity from a voltaic pile. 1890 Garuti electrolytic system tried at Brera Palace, Milan. Art of obtaining metals from solutions by galvanic action *disc.* M. H. Jacobi 1801–). *See also* Electro-plating.

Electrolytic Dissociation (chemistry) 1886 Theory *prop.* by Svante Arrhenius (Swed) (1859–1927).

Electrometer, Torsion *Inv.* by Charles Augustin de Coulomb (Fr) (1738–1806).

Electro-plating 1800. Johann Wilhelm Ritter (Ger) (1776–1810) *disc.* electro-plating of copper. 1836. Electro-plating *intro.* and *pat.* by G. R. and H. Elkington, who plated copper and brass with silver and gold. 1842. Ruolz electro-plating process *pat.* in England by Pilkington Brothers and Christofle, of Paris. 1843. Böttger *inv.* electro-platinizing process. 1875. Böttger proposed use of double sulphate of nickel and ammonia for nickel-plating. 1878. Planzan added citric acid to Böttger's process. 1879. Weston *inv.* the boric acid process of nickel-plating. 1897. Förster *dev.* process of depositing nickel from hot solution of nickel sulphate.

Electro-magnet 1820. Electro-magnet *disc.* Hans Christian Œrsted (Dan) (1777–1851). 1820. A. M. Ampère *disc.* magnetic effect of electric current passing through a coil of wire. 1832. Michael Faraday (1791–1867) *inv.* copper disc and magnet electric generator. (*See also* Generator, Electric.)

Electro-magnetic Waves *See* Waves, Electro-magnetic.

Electrometer *Inv.* Count Alessandro Volta (It) (1745–1827).

Electronic Charge 1897. First measured by Townsend.

Electron Word first used by Johnstone Stoney to designate an electric charge of atoms. (Now used to describe an alpha particle.) 1906. Electron *disc.* J. J. Thomson (1856–1940).

Electron Charge. Determined by Robert Andrews Millikan (U.S.) (1868–1953).

Electron, Intrinsic angular momentum of 1924–25. Conception *prop.* by Goldsmit and Ulenbeck.

Electron Diffraction Pattern 1925. *Disc.* C. J. Davidson and Gerner, whilst experimenting with crystallized nickel.

Electrophonoscope 1890. First shown at South Kensington, London.

Electrophorus *Inv.* Count Alessandro Volta (It) (1745–1827).

Electroscope Simple form of *inv.* William Gilbert (1544–1603). Linen thread electroscope made by Benjamin Franklin (U.S.) (1706–90). Pith ball electroscope used by C. F. de C. Fay, Canton and Henley. 1787. Gold leaf electroscope *dev.* by Abraham Bennet.

Electrotyping 1799. *Disc.* by Count Alessandro Volta (It) (1745–1827). 1800. Electrotyping *intro.* into England. 1805. Brugnatelli (It) gilded silver coins. (*See* Electro-plating.) 1841. Electrotyping printing *intro.* into England. 1841. De la Rive (Fr) gilded brass articles. 1873. Carl Gustav Jacob Jacobi (Rus) (1804–51) electrotyped medals.

Elements, Periodic Tables of 1815. William Prout suggests that all elements are composed of one simple element—hydrogen. 1865. William Odling proposed a Periodic Table, as did Sir John Alexander Newlands, in his "Law Octaves." 1869. Dimitri Mendeléev *pub.* his Periodic Table. 1870. Lothar Meyer proposed a system.

Elevator *c.* 200 B.C. Manpowered drum-type elevator *desc.* by Archimedes. 1573. Rocking elevator ("Ladder of Volturius," "Lazy-tongs," or "Nuremberg Scissors") used by Juanelo Tuviano (Sp) for supply of water to Alcazar Palace, Toledo (250 ft. lift and 2,000 ft. run). 1655 "Lazy-tongs" elevator *desc.* Marquis of Worcester for tobacco tongs: No. 49 in his *Century of Inventions.* 1850. Drum-type elevator *inv.* 1854. Elisha Otis *inv.* safety elevator. *See also* Conveyors and Lifts.

Elevator, "Sack" 1850. *Inv.* by Thomas Moore Sack, of Belfast.

ELSHOLTZ, Johann Sigismund (Ger) (1623–88) 1676. *Obs.* thermo-luminescence of heated fluorspar.

Embroidering Machine 1834. 20-needle embroidering machine *inv.* Joshua Heilmann, of Mulhouse, Alsace.

Embriology 1828. Von Baer (Ger) (1792–1876) founded science of embriology and traced all development stages of an animal from its first appearance in the egg to its final birth.

Emery, Artificial 1842. *Inv.* Henry Barclay.

Emery-paper 1843. *Inv.* R. Edwards.

Enamelling 1200 B.C. Earliest record of fused enamel ware. 1545. Bernard Palissy made his first enamelled ware. 1799. Dr. Hinkling *inv.* enamelling process for saucepans. 1839. Clarke *pat.* new method for enamelling kitchenware.

ENCKE, Johann Franz (Ger) (1791–1865) 1822. Determined parallax of sun to be 8.57 seconds.

Energy 1854. Term *intro.* by William Thompson, Lord Kelvin (1824–1907).

Energy, Conservation of 1842. Law of *disc.* by Julius Robert von Mayer (Ger) (1814–78). 1847. First referred to by Hermann von Helmholtz (Ger) (1821–94).

Energy, Dissipation of 1824. Axiom enunciated by Sadi Carnot (Fr) (1796–1832).

Engine, Electric reciprocating *c.* 1860. *Inv.* by Froment. Worked with a solenoid-driven beam.

Engine, Free-piston *Inv.* Pescara (Fr).

Engine, External Combustion (Gas) 1820. Rev. E. Cecil *des.* external combustion engine. 1823–26. Samuel Brown *pat.* external combustion pumping engine. 1827. External combustion engine-driven boat used on River Thames. 1832. Brown external combustion pumping engines in use at Croydon, Brompton (London), and Soham, Cambs.

Engine, Gas 1799. Phillipe Lebon (1769–1804) (Fr), of Brachay, *pat.* coal-gas-engine which was improved, but un-

perfected by 1804, when he was assassinated. 1807. De Rivaz (Fr) *pat.* "gas-driven machines." 1823–26. Samuel Brown *pats.* external combustion gas engine pump. 1827. Brown made boat fitted with gas engine which was tried on River Thames, and in 1835 fitted a road carriage. (*See also* Engine, internal-combustion.) 1838. William Barnett (U.S.) made twin-cylinder gas engine pump with separate combustion chamber. 1839. Wright (U.S.) *inv.* gas engine with double-acting cylinder, gas-jet fired. 1839. Johnston (U.S.) *inv.* gas engine using hydrogen and oxygen. 1844. John Reynolds proposed using hot-wire ignition with a primary battery. 1850. Stéphard (Fr) suggested use of magneto-electric machine for gas engine ignition. 1855. Dr. Alfred Drake, of Philadelphia, U.S., *des.* a gas engine with hot-point ignition, which worked successfully at the Crystal Palace Exhibition, New York. 1855. 1857–58. Dégrand (Fr) *pat.* gas engine in which gas mixture was compressed in the working cylinder. An important step in gas engine history. 1861. Pierre Hugon (Fr) *pat.* gas engine with flame ignition and water-cooling, which proved better than Étienne Lenoir's of 1860. 1862. Beau de Rochas (Fr) *pat.* four-cycle system for internal-combustion engines, but never made a machine. 1864 (Dec.). 143 Lenoir engines working in Paris and one in London at Messrs. G. B. Kent, Great Marlborough Street. 1867. Dr. Nicholas Otto (Ger)

(1832–91) and Eugéne Langen (Ger) produced a very noisy vertical gas engine exhibited at the International Exhibition. It was based on Abbe de Hautfeuilie's "gunpowder" engine of 1678. 1873. Brayton (U.S.) *des.* a gas engine which was later *dev.* into a petrol engine. 1881. Two-stroke gas engine *inv.* Sir Dugald Clerk. 1883. Gottleib Daimler (Ger) (1834–90) *intro.* light oil engine. (*See* Internal combustion engine, Motor car, etc.) 1886. Daimler and Karl Benz (Ger) (1844–1929) produced gas engine-powered road vehicles. 1883. Griffin *intro.* gas engine with cycle involving two scavenging strokes. 1885. Atkinson *intro.* gas engine working on differential principle, with two pistons. 1906. H. A. Humphrey, collaborating with W. J. Randall *inv.* a gas engine pump in which the piston was replaced by an oscillating water-column. *See also* Engine, Internal Combustion.

Engine, Hot-air 1807. First *inv.* by Sir George Cayley (1774–1857). Used hot furnace gases. 1816. Rev. Dr. Stirling *pat.* closed cycle hot-air engine with extra compression cylinder. 1830. John Ericsson (Swed) *inv.* hot-air engine cycle—his "caloric engine." 1837. Caley's hot-air engine *pat.* 5 h.p. obtained from 20 lb. of coke. 1843. Stirling hot-air engine of 45 h.p. installed at a Dundee foundry. 1872. Ericsson made unsuccessful solar-driven hot-air engine. (*See* Solar energy.) 1875. Robinson made hot-air engines of one and two manpower.

1875. Heinrici popularized hot-air engines on the Continent. 1876. Rider (U.S.) *inv.* hot-air engine. 1880. Lehmann *dev.* very large hot-air engines.

Engine, Internal combustion 1678. Abbé de Hautfeuille proposed an engine with gunpowder as motive-power. 1680. Christiaan Huygens *desc.* engine with gunpowder as powersource. 1791. John Barber *pat.* gas-turbine engine (*q.v.*). 1794. Robert Street *pat.* internal combustion engine. 1799. Philip Lebon, of Brachay, France *pat.* engine. 1826. Capt. Samuel Morey (U.S.) *pat.* internal combustion engine with electric ignition, carburetter, poppet-valves, and water-cooling. 1838. William Barnett *pat.* first internal combustion engine in which ignition was effected from outside the cylinder. 1844. John Reynolds *inv.* platinum-wire ignition with a battery. 1850. Stéphard used a magneto-machine for ignition instead of a battery. 1855. Dr. Drake *inv.* "hot-spot" ignition. 1857. Barsanti and Matteucci *desc.* a gas engine the *des.* of which was materially the same as that of Otto and Langen (1867). 1858–59. Dégrand *pat.* gas engine in which compression took place in the cylinder itself. 1860. Étienne Lenoir *pat.* his gas engine. 1858. Hugon (Fr) *pat.* 2½ h.p. gas engine with gas-jet ignition. 1860. J. H. Johnson, of Lincoln's Inn Fields, London *inv.* gas engine with electric ignition. 1861. Lenoir's engine modified by Kinsey and Kinder. 1862. Beau de Rochas (Fr) evolves the "four-cycle" principle—later known as the "Otto Cycle." 1867. Otto and Langen produce a rough and noisy gas motor. 1878. Dr. Otto *intro.* his famous gas engine. *c.* 1880. Bisschop, Simon and Ravel gas engines *intro.* 1881. Dugald Clerk *inv.* two-stroke gas engine. 1883. Griffin *intro.* engine with a six-stroke, scavenging cycle. 1885. Atkinson *intro.* twin-piston, single-cylinder, "differential" gas engine. *c.* 1875. Brayton (U.S.) gas engine of 1873 converted by its designer to run on petroleum fuel. 1883. Gottleib Daimler originated a small, high-speed engine consuming the light oil now known as petrol. 1886. Preistman *intro.* first practical oil engine, which embodied Édouard Etève's *pat.* of 1884. 1890. H. Stuart Ackroyd *inv.* the Hornsby-Ackroyd oil engine, the first to operate on the compression-ignition principle. 1895. Dr. Rudolf Diesel *intro.* his four-cycle oil engine. (*See* Diesel rail-car and Diesel ships.) 1884. Karl Benz (Ger) (1844–1929) *prod.* his first internal combustion engine independently of Daimler. (*See* Motorcar.) 1927. Diesel-steam locomotive engine *inv.* by W. J. Still and Lt. Col. E. Kitson-Clark. 1930. Automatic diesel engine evolved by Cummins (U.S.). *See also* Engine, Gas.

Engine, Internal combustion rotary *See Pats.* of André Beetz, Chaudun, Dodement, Gardner-Sanderson and Vernet.

Engine, Naptha 1888. *Inv.* Escher Wyss (Swit).

Engine, Steam 1663. Marquis of Worcester *des.* atmospheric engine. 1678. Robert Hooke *pat.* steam engine. 1682. Sir Samuel Morland *inv.* pumping steam engine. 1685. Denis Papin exhibits model pumping steam engine. 1698. Capt. Thomas Savery (1650–1715) *pat.* steam engine. 1705. Dartmouth, Devon plumber Thomas Newcomen and glazier John Cawley *pat.* first practical model atmospheric steam engine. 1707. Papin produced full-sized steam engine. 1712. Newcomen erects his first full-sized steam engine at Dudley Castle. 1718. Henry Beighton, of Newcastle upon Tyne *inv.* self-working valve-gear. 1723. Leupold *des.* two-cylinder, expansively operated steam engine. 1765. James Watt builds his first steam engine. 1766. Blakey *pat.* improvements to Savery's engine. 1769. Watt perfects his steam condenser. 1774. John Smeaton builds 74½ h.p. Newcomen engine at Wheal Busy, Chacewater, Cornwall. Watt's first two engines installed to drain a colliery and blow a blast-furnace. 1781. Jabez Carter Hornblower (1744–1815) *inv.* two-cylinder compound steam engine. 1781. Watt *pat.* rotary steam engine and used it (1785) to drive a cotton-mill. 1782. Watt *pat.* double-acting cylinder and use of steam expansively. 1785. William Murdock (1754–1834) *inv.* oscillating-cylinder engine. 1802. Murdock *des.* rotary engine of ½ h.p. 1804. Arthur Woolfe *intro.* compound engine as a modification of a Watt engine. 1823. Perkins *inv.* engine working at 500 p.s.i. 1824. J. P. Allaire (U.S.) *inv.* compound engine. 1827. Perkins increases pressure to 800 p.s.i., expanding steam eight times. 1830, James Dakyn; 1833, John Ericsson; 1834, Earl of Dundonald; 1835, Reynolds and Avery; 1836, Yule; 1842 Lamb; and 1847, Behrens; *inv.* various forms of rotary steam engines. 1849. George Henry Corliss (U.S.) (1817–88) *inv.* the valve-gear bearing his name. 1856. Compound steam engine *intro.* on ships by John Elder. 1857. Modern rotative form of steam engine *inv.* by John McNaught and Edward Cowper. 1863. First triple-expansion ship engine *des.* by A. C. Kirk. 1866. Superheated steam engine *intro.* by E. Danford (U.S.). 1871. Three-cylinder steam engine *intro.* by Peter Brotherhood (1838–1902) and J. Krig. 1874. P. W. Willans *inv.* high-speed steam engine. 1880. Willans produced high-speed, central-valve engines up to 2,400 h.p. 1890. Bellis and Morcom *pat.* high-speed engine running up to 800 r.p.m. 1905. "Pioneer" steam bus driven by a rotary engine appears in London. 1910. Five-cylinder rotary engine *des.* by Leslie Walker. 1911. A. G. M. Mitchell (Australia) *des.* eight-cylinder swash-plate steam engine for the Australia Gas Light Company, of Sydney. (*See also* Locomotive, Steam Railway and Road Vehicle, Steam, and Engine, Steam Traction.)

Engine, Two-fluid-cycle 1913. Two-fluid-cycle engine, using mercury and water, *inv.* W. L. R. Emmett (U.S.).

Engraving 1491 B.C. Mentioned Exodus xxviii. 11. 1120. B.C. Practised by Chinese. 1346 (A.D.). Earliest known line engraving. 1423. Earliest known line engraving in England. 1461. Earliest known copper engraving in England. 1486. Art of cross-hatching *inv.* Michael Wohlgemuth (Ger). 1493. Wood-block engraving used in Nuremberg Chronicle news-sheet. 1476. Wood-block printing blocks *intro.* by William Caxton. 1513. Engraved wood printing-block used in England in news pamphlet re Battle of Flodden Field. 1773. Wood-block engraving revived and perfected by Berwick. (*See also* Mezzotint, Aquatint, Lithography, etc.)

Envelope (postal) *c.* 1726. *Intro.* in France. 1840 (May 1). *Intro.* in England with embossed stamp and design by Irish artist William Mulready (1786–1863). 1844. envelope-making machine *inv.* George Wilson. *c.* 1845. Envelopes superceded folded paper sheet and sealing-wax for postal packets in England.

Enzymes 1878. Term first used by Willy Kühne (Ger) (1837–1900).

Epicycloid 1525. Geometrical form *disc.*

Epidermis *Disc.* by Dr. Marcellus Malpighi (It) (1628–94).

Epsom Salts 1618. Mineral spring at Epsom, Surrey *disc.* 1695. Epsom salts first prepared from Epsom springs by Nehemiah Grew and called by him "sal Anglicum" and "bitter salt."

Equinoxes, Precession of the *Disc.* by Hipparchus (160–145 B.C.) as being 25,866 years.

ERATOSTHENES (276–196 B.C.) *Disc.* obliquity of the ecliptic and *inv.* Armillary spheres.

Erbium (element) 1839. *Disc.* by Mosander.

Ergot 1954. Active principle of lysergic acid synthesized by B. B. Woodward, of Boston, U.S.

Escapement, Watch and Clock 1671. William Clement (or Robert Hooke) *inv.* anchor escapement. 1675. Christiaan Huygens (Robert Hooke, or Abbé Jean de Hautefeuille) *des.* spring balance wheel. 1715. Graham *inv.* dead-beat escapement. 1720. Cylinder escapement *inv.* George Graham (1673–1751). 1852. Sir Edward Beckett (later Lord Grimthorpe) *inv.* gravity escapement in clock "Big Ben," Westminster, London. (*See also* Watch and Clock.)

Esperanto *Inv.* Dr. Zamenhof (1859–1917).

Ester Gum 1884. First produced from rosin by E. Schaal.

Ether (sulphuric ether) 1540. Valerius Cordus *disc.* method of making ether, naming it sweet oil of vitriol. 1729. Frobinius *disc.* various properties of ether. 1818. Michael Faraday (1791–1867) *disc.* pain-dulling effect of inhaling ether. 1822. Effect again *obs.* by U.S. surgeon Godman and in 1832 by Dr. Mitchell, of New York; who both experi-

mented with it as an anaesthetic. 1842. Ether first used on a patient by U.S. surgeon Jefferson, of Georgia. 1846 (Sept. 30). Ether first administered as a dental anaesthetic by W. T. G. Martin, of Boston, Mass. 1846. Edinburgh surgeon Liston performed internal operation with ether at University College Hospital, London (Dec. 21); when ether administered by Dr. Squire. 1847. Ether adopted as anaesthetic by Dr. Nikolai Pirogoff (Rus) (1810–81).

"Etherphone," or "Thereminvox" 1924. Radio-electric musical instrument *inv.* Leo Theremin.

EUCLID of Alexandria (*c.* 300 B.C.) First gave a definite form to systematic exposition of mathematics.

Eudiometer Various types *inv.* by Joseph Priestley, Alessandro Volta, Édouard Séguin and Claude-Louis Berthollet.

EULER, Leonhard (Ger) (1707–83) 1736. Wrote *Mechanics, or the Science of Motion created Analytically*. The first textbook on methematics. 1760. Devised early gearing techniques (*q.v.*).

Europium (element) 1889. *Disc.* by Sir William Crookes (1832–1919).

Eustacian Tubes *Desc. c.* 500 B.C. by Alemdeon (Gr), of Croton, South Italy. Re-*disc.* by Bartolomeo Eustacio (1520–74). (*Disc.* not *pub.* until 1714.)

Evacuator (surgical) *Inv.* Henry Jacob Bigelow (1810–90), of Boston, Mass.

EVANS, Oliver (U.S.) (1755–1819) 1794. *Des.* continuous flour-mill (*q.v.*). 1805. *Inv.* first self-propelled road vehicle to run in U.S.

Evolution Theory 1842. *Prop.* by Charles Darwin (1809–82). 1859. Theory of natural selection enunciated by him.

Excavators 1655. One suggested by Marquis of Worcester in his *Century of Inventions*, No. 92. 1726. François Joseph Camus (Fr) *inv.* mechanical shovel with all the essential features of modern machine. *c.* 1830. Steam engine brought into use in U.S. during *const.* of first Pacific Railway. 1865. Excavator machine *inv.* by Frey Freres et Sayn (Fr) with horizontal area of 36 ft. and

lift of 11 ft. 6 in. 1884. Mm. Weyher and Richmond, of Pantin, France, *des.* machines for use on Panama Canal.

Exclusion Principle (nuclear physics) *Disc.* by Pauli.

Extensiometer 1856. Electric extensiometer (strain-gauge) *inv.* Lord Kelvin (William Thomson) (1824–1907). *See* Tensil strain.

EYDE, Samuel (1866–1940) 1903. With K. Birkeland pioneered electrolytic nitrogen fixation process.

Eye Effect of ciliary muscle on the lense of the eye *disc.* Thomas Young (1773–1829), of Taunton, England.

F

Fabrics, Crease-resistant 1929. *Pat.* granted to Messrs. Boffey, Foulds, Marsh and Tankard, of Tootal, Broadhurst Lee, Co.

FABRIZZI, Hieronymo (It) (1537–1619) Applied mechanical principles to anatomy of muscular movement.

Facsimile Transmission 1847. Record of first *Pat.* relating to.

Factor Theorem 1631. *Disc.* Bonaventura Cavalieri (It) (1598–1647).

FAHRENHEIT, Daniel Gabriell (Ger) (1686–1736)

Made thermometers in Holland. 1721. *Disc.* phenomenon of super-cooling water.

FAIRBAIRN, William (1789–1874) 1844. *Inv.* Lancashire boiler.

Fallopian Tubes (anatomy) *Disc.* by Gabriele Fallopio, of Padua (1523–62).

Fan 40 B.C. Crank-handled winnowing fan used in China. 180 (A.D.). Rotary ventilating fan used in China. 1572 Hand fan *intro.* England from France. 1837. William Fourness, of Leeds *inv.* exhaust fan and inaugurated modern method of mine ventilation. 1848. George Lloyd *pat.* centrifugal mine fan. Guibal *inv.* rotary fan and produced one 17 ft. in diameter.

FARADAY, Michael (1791–1867) Responsible for many electrical and chemical *disc.*—electro-magnetic induction, butylene, etc. (*See* individual entries.)

Fatigue-testing Machine Pre-1858. First *inv.* A. Wohler (Ger) of Hanover. *See also* Extensiometer, Tensile strength.

Fats, Animal 1823. Michael Eugène Chevreul (Fr) (1786–1889) *disc.* animal fats to be compounds of glycerine and aliphatic or organic acids.

Fats, Hydronization of 1901. *Inv.* Dr. W. Normann (Ger) (*b.* 1870)—a process of solidification.

FAURE, C. A. (Fr) 1881. *Inv.* electrical accumulator.

FAVRE, Louis (Fr) of Geneva 1780. *Inv.* cylinder musical-box.

Felt (textile) 1813. First made from shoddy at Botley, Yorks.

Fermium (element) 1952. *Disc.* Glen T. Seaborg and A. G. Liorso (U.S.).

Ferns, Asexuality of 1628. *Disc.* by an Italian author in Lincei Academy.

Ferro-electricity 1921. *Disc.* Valasek.

Fertilizers (horticultural) 1602. Peruvian guano used as fertilizer in Portugal. 1804. Theo de Saussaur (Fr) *disc.* saltpetre promoted growth of cereals. 1830. Nitrates first shipped from Peru and Chile. 1843. John Bennet Lawes and J. H. Gilbert *disc.* nitrate of potassium and potash were most needed in agriculture. *c.* 1840. Peruvian guano first used in England. 1852. Potash found by drilling at Strassfurt. 1861. Basic slag first used as source of phosphorus. *c.* 1868. Lawes proved superphosphate was a good fertilizer. 1880. Bones first used as fertilizer. *See also* Fumigants and Insecticides.

Feynman Theory (nuclear physics) 1948. *Prop.* by Richard P. Feynman (U.S.). *See also* Neutrino particle.

File 1093 B.C. Mentioned 1 Samuel xiii. 20–1. *c.* 1000 B.C. Mentioned by Homer. 1495. Leonardo da Vinci sketched machine for making files, with cam, gearing and cutting tools. 1697. Earliest *inv.* for cutting files. 1859. File-making commenced at Leeds.

Filter 1791. Sand filter *inv.* James Peacock. Early 18th cent. First mention of woollen cloth filter for sugar refining. 1812. Cylinder filter *inv.* Paul, of Geneva. 1814. Multiple sand filter *inv.* Ducommon (Fr). 1815. Pressure filter *inv.* Count Réal (Fr). 1819. Compressed-air filter *inv.* Hoffman, of Leipsig. 1824. Bag, or stocking filter *inv.* Cleland. 1831. Earthenware ascending filter *inv.* Lelogé (Fr). 1845. Howard *inv.* linen filter. 1853. Dr. Stonehouse *inv.* charcoal air filter. 1827. First slow sand drinking-water filter *inv.* James Simpson. 1791. Unglazed earthenware filter *pat.* 1868. Filter-pump *inv.* Robert Wilhelm Bunsen (1811–99). 1886. Kieselguhr filter *pat.* Heddle and Stewart (also by Weischmann (U.S.). 1904. Gottleib Daimler (Ger)

(1834–1900) fitted gauze petrol-filter to Phoenix-Daimler motor car engine. 1919. Stack-type filter *inv.* R. W. Bunsen (Ger).

Fire *c.* 500,000 B.C. Pekin man used fire in Asia. 500,000–235,000 B.C. Chellean (Stone Age) men in Europe used fire for heating, lighting and protection. 50,000–10,000 B.C. Late Stone Age man in Europe made fire by striking flint with iron pyrites. 11th–12th cent. (A.D.). Marcus Græcus wrote about incendiary compositions in his *Book of Fires. See also* Fireworks.

Firearms, Hand 1331. Hand firearms used by Germans at battle of Vicidale, Italy. 1381. Earliest mention of hand firearms (fire-sticks, or batons á feu). 1398 (post). Earliest hand firearms to have separate chambers for powder and bullet. *c* 1425. First "firelock" or "matchlock" *intro.* with standing aim. 1435. Hand-grenades first used. 1515. Wheel-lock *inv.* 1522. Spaniards used matchlock hand-gun (arquebus) at Bicocca, Italy. 1540. Spaniards *intro.* large musket on stand. 1540. First flintlock pistol made at Pistoia, Italy. 1635. Form of flintlock musket *inv.* 1704. Breech-loading musket *inv.* De la Chaumette (Fr). Breech-loading carbine *inv.* Maurice de Saxe (Fr). 1785. N. le Blanc (1742–1806) made muskets with interchangeable parts. (*See also* Rifle and Rifling.)

Fire-blower, Steam (sufflator) 1405. Pictured by Konrad Kyeser with bronze nozzle, as used by soldiers at camp-fires. 1495. Leonardo da Vinci sketched bronze-nozzled bellows. *See also* Bellows.

Fire-engine 250 B.C. Mentioned by Pliny. 1663. "Modern" fire-engine *inv.* Van der Heyden (Hol). 1682. Hydraulic fire-engine *inv.* 1720. Air pressure chamber for fire-engines *inv..* 1721. R. Newsham *pat.* fire-engine. 1792. Charles Simpson *imp.* Newsham's fire-engine. 1793. Joseph Bramah *inv.* reciprocating fire-engine fitted with Barton's pistons (*q.v.*). 1830–39. Steam fire-engine *inv.* Braithwaite and Ericsson, London. 1840. Steam fire-engine *inv.* Paul Rapsey Hodge, of New York. 1895. First motor fire-engine *dem.* at Tunbridge Wells, Kent.

Fire Escape 1766 and 1773. *Pat.* in France. 1809. First British *pat.* Davis, of London.

Fire Extinguisher 1734. M. Fuchs (Ger) *inv.* fire extinguisher with filled glass balls to throw into fire. 1761. Zachary Grey used balls filled with sal ammoniac. 1762. Dr. Godfrey, of London used sal ammoniac filled balls burst by gunpowder. 1792. Von Ahen (Swed) and Nils Moshein (Swed) independently *inv.* fire extinguishers using chemicals and water.

Fire Piston (ignition device) 1807. Fire piston contained in walking-stick *pat.* in England by Richard Lorenz. *See* also Diesel engine.

Fireworks 7th cent. (early). Chinese books *desc.* use of fireworks. 1360. Re-*inv.* in Florence. 1588. First fireworks display. 1697. First firework display in England to celebrate Peace of Ryswick. 1871. First firework display at Crystal Palace, Sydenham, London, by Brock.

FISCHER, Emil (Ger) (1852–1919) 1887. *Isol.* isometric hexose sugars. 1897–99. Synthesized caffeine and theobromine.

Fishplate, Railway 1843. *Inv.* in U.S. 1847. *Imp.* by William Bridges Adams in England jointly with Robert Richardson.

Fishing-reel 13th cent. Appears in Armenian manuscripts. 14th cent. In use in China. 1641. First mention of in Europe.

FIZEAU, Hippolyte Armand-Louis (Fr) (1819–96) 1849. Performed many important experiments in light and optics.

Flag Code, International 1857. *Dev.* Sir Home Riggs-Popham.

Flageolet 1803. *Inv.* William Bainbridge.

Flail Early 5th cent. Jointed flail first mentioned by St. Jerome.

Flame, Manometric *Inv.* Samuel Koenig (Ger) (1712–57).

Flame, Sensitive 1777. *Disc.* by Huggins. 1858 Re-*disc.* and *obs.* by Prof. Leconte (U.S.) and *dev.* into the pyrophone, or flame organ. 1869. "Lustre chantant" *dev.* by F. V. Kastner, of Paris.

FLAMSTEED, John (1646–1719) Compiled earliest star catalogues.

Flats (housing) 1830. System of living in flats (Fourierism) devised by Charles Fourier (Fr) (*d.* 1837). He planned flats,

named phalansteries, for 400 families in one building. The scheme failed.

Flax 3000 B.C. Growing established in Near East. 1832. Heckling machine *inv.* Phillipe de Girard (Fr). 1850. Le Chevalier P. Claussen (Fr) *inv.* method of making short-staple flax and mixed flax with cotton and woollen goods.

FLEMING, Sir Alexander (1881–1955) 1904. *Inv.* thermionic valve.

Flow-meter 1739. *Inv.* Henri Pitot (Fr). *See also* Venturi tube.

Fluid Flywheel *See* Hydraulic coupling and Power transmissions.

Fluid Flow 1910. Ludvig Prandtl (Ger), of Hanover made paddle-wheel test tank to check fluid flow.

Fluorescence Word *intro.* by Sir G. G. Stopes.

Fluorine (element) 1771. Carl Wilhelm Scheele (Ger) (1742–86) obtained free hydrufluoric acid gas. 1818. Sir Humphry Davy (1778–1829) attempted to produce free fluorine. 1841. Knox Brothers. attempted to *isol.* fluorine, as did Frémey (Fr) (1856), Kammerer (1865), and Finkener (1867); all were abortive. 1886. Fluorine finally *isol.* by Henri Moissan (Fr) (1852–1907).

Fluorspar 1838. Fluorescence of *disc.* by Sir David Brewster.

Flute 1823. New system of keying *inv.* Theo Boehm.

Fluxions 1665. Doctrine *prop.* by Sir Isaac Newton (1642–1727). *See also* Differential calculus.

Flying-bomb *See* Aircraft, pilotless.

Fly-press 1651. One in use at British Royal Mint. 1790. New type *inv.* by Matthew Boulton (1728–1809) for use at the Royal Mint.

Fly-shuttle 1738. *inv.* by John Kay (1704–64).

Flywheel Date of *intro.* unknown. 1430. Recorded application to crank and connecting-rod by German millwright. Early 16th cent. Books show flywheels or weights at end of rotating arms were in use, especially for crank-driven pumps.

Focal-plane shutter (camera) 1861. *Inv.*, together with roller-blind shutter, by William England.

Fodder-cutting Machines (chaff-cutters) 1731. Earliest *inv.* by Thomas Ryley and John Beaumont. 1770. James Edgill *inv.* spiral-knifed machine, and James Sharp *inv.* bean-splitting mill and winnowing machine (*q.v.*). 1787. James Cooke *des.* modern type "chaff-cutter" with spiral knives mounted on spokes of operating wheel.

Folic Acid 1941. *Disc.*

Folinic Acid 1948. Prepared by Live and associates.

FOND, Jean-René Sigaud de la (Fr) Assisted in the early *dev.* of static electricity producing machines.

Food Preservation, Methods of 1795. Nicholas Appert (Fr) of Paris, preserved food by boiling and sealing in glass bottles. 1810. Appert parboiled and Donkin boiled foodstuffs, sealing them in tinned-iron cans and sealing a soldered hole therein. (Canned storage in Britain by Peter Durand.)

See also Tinned iron. 1824. Gamble preserved food in tins for the Arctic Expedition of that year. 1842. Bevan *inv.* the vacuum chamber, and Dirchoff (Rus) *inv.* milk powder. 1850. Dried meat biscuit *inv.* 1851. M. Mason (Fr) *inv.* method of dessicating vegetables. *See also* Canning, Curing, Bottling.

Forceps (surgical) Obstetric forceps *inv.* by Hugenot refugee Peter Chamberlain the Elder (1560–1631) and *inv.* kept secret for 150 years. Axis traction handles added to Chamberlain's forceps by Étienne Stéphane Tarnier (Fr) (1828–97). 1720. Spoon forceps *inv.* Jean Palfyn (Fr). 1733 French-lock forceps *inv.* by Dusée (Fr). 1733. Groove-locked forceps *inv.* Edmund Chapman. 1745. Wooden forceps *inv.* William Smellie (English-lock type). 1751. Screw-handled forceps *inv.* by John Burton, of York. Bone-cutting forceps *inv.* by Robert Liston (1794—1847). "Spencer-Wells" (haemostatic) forceps *inv.* Sir Thomas S. Wells (1818–97).

Forces, Parallelogram of 1725. Significance of first recognized by Pierre Varignon (Fr).

FORD, Henry (1863–1947) Evolved and *dev.* mass-production methods and applied them to making motor-cars.

Forks (cutlery) 1608. Three-pronged fork *intro.* into England from Italy by Thomas Coryat. *c.* 1610. Two-pronged steel forks first made at Sheffield.

Formaldehyde 1856. *Disc.* by Debus.

Formolite (resin) 1903. *Disc.* Nastyukov (Rus).

Formulæ (chemical) 1811. Fundamental method expounded by Amadeo Avogardo (It) (1776–1856). 1858. Avogardo's method revived by chemist Stanislao Cannizarro (It) (1826–1910).

FORTIN, Jean (Fr) (1750–1831) Engineer responsible for early *dev.* of the lathe.

FOUCAULT, Jean-Baptiste-Leon (Fr) (1819–68) *Invs.* and *discs.* relating to arc-lamps, velocity of light, gyroscope, and the proving of the earth's diurnal rotation by a free-swinging pendulum. (All of which *q.v.*)

FOURIER, Jean Baptiste-Joseph (Fr) (1768–1830) 1822. *Prop.* theory of heat.

FOURNEYRON, Benoit (Fr) (1802–67) *Des.* water-turbines. Was pupil of Prof. Claude Burdin, who *des.* original enclosed, submerged, horizontal water-wheel, which he named a turbine (*q.v.*).

FOX-TALBOT, W. H. (1800–77) Born Melbury, Dorsetshire. *Inv.* callotype process by which any number of photographic prints could be made from one negative.

FRANCIS, James B. (1815–92) Inventor of the inward-flow water-turbine design (*q.v.*).

FRACASTORO, Hieronymo, of Verona (1484–1553) First *isol.* different types of fever and had earliest ideas of infection ("seeds," or germs).

Francium (element) 1939. *Disc.* by Mmle. Margaret Perey (Fr).

FRANKLAND, Sir Edward (1825–99) 1849–53. Devised new principles of organic chemical analysis and synthesis.

FRANKLIN, Benjamin (U.S.) (1706–90) 1752. *Inv.* the lightning conductor. Experimental chemist and physicist.

FRAUNHOFER, Joseph von (Ger) (1787–1826) Physicist. 1814–17. *Disc.* the element-indicating lines named after him on the spectrum.

Fraunhofer Lines *See* previous item.

Free Energy, Debye 1923. *Disc.* by Peter J. W. Debye (Hol).

Freezing, Physical Law of 1788. Blagden formulated law relating to freezing-point of solutions (cryoscopy). 1861. Rudorff announced Blagden's law as his own new *disc.*

Freon (Dichlorodifluoromethane) (refrigerant) 1930–31. *Disc.* as a non-toxic refrigerant by Thomas Midgley Jnr. (U.S.). *See also* Tetraethyl lead. *See also* Cryogenics, Refrigerants, Refrigerators.

FRESENIUS, Karl Remigius (Ger) (1818–97) Chemist. *Dev.* methods of chemical analysis.

FRESNEL, Augustin-Jean (Fr) (1788–1827) *Prop.* the undulatory theory of light and proved it by subtle experiments.

FRIEDEL, Charles (1832–99) Devised *imp.* methods of chemical analysis and synthesis.

Friction, Knowledge of Expressed by Leonardo da Vinci (1452–1519). 1699. Laws relating to friction devised by G. Amontons (1663–1705).

FRIESE-GREENE, William (1855–1921) 1889. *Inv.* the cinematograph (*q.v.*) in England.

FRISIUS, Rainer Gemma (Hol) 1544. Used camera obscura to *obs.* solar eclipse of Jan. 25. 1544.

Front-wheel Drive (automobile engineering) 1769. Nicholas Joseph Cugnot (Fr) (1725–1804) *des.* and *const.* the first full-sized mechanically operated road vehicle, which had front-wheel drive. 1895. The "Electrobat" electric vehicle had front-wheel drive, as had the "Christie" motor-car of 1904. *See also* Motor Car.

FROUDE, William (1810–79) 1877. *Inv.* the hydraulic dynamometer.

FUCHS, Leonides (Ger) (1501–66) Pharmaceutical chemist who *disc.* the fuschia in Mexico, 1542. (The flower was *intro.* into England in 1830.)

Fuel Cell 1952. F. T. Bacon *pat.* oxy-hydrogen fuel cell which gave 650 milliamps of electric current at 0.79 volt per sq. cm.

Fuel Injection 1904. Fitted on Pope motor-car. *See also* Diesel engine.

Fuel, Pulverized 1807. *Inv.* in France for early internal combustion engine.

FULBERT (*c.* 1460–1527) Mathematician and scientist in the cathedral school of Chartres.

Fulling *c.* 730 (A.D.). Process *inv.* by Nicias, Roman governor of Greece during the Roman occupation (Pliny). 17th cent. (turn of). Water-driven fulling mills in common use.

FULTON, Robert (U.S.) (1765–1815) 1807. *Des.* and *const.* steam-driven paddle-wheel ship *Fulton*, which made 150 mile trip up the Hudson River.

Fumigants (agricultural) 1920. First used. *See also* Insecticides.

FUNK, Casimir (*b.* 1884) 1912. Obtained crystalline substance from rice-polishings and named it "Vitamine."

Furnace, Iron-smelting Copula furnace *inv.* John Wilkinson (1714–72). 1784. Revaberatory furnace (puddling) *inv.* Henry Cort. *See also* Smelting, Iron. 1873. Siemens-Martin open-hearth furnace *intro.* 1879. Pourcel and Valrand variant of open-hearth furnace *intro.*

Furnace, Electric 1879. First electric furnace (arc) *inv.* by Clerc. 1882. Electric furnace *inv.* by Ernst Werner von Siemens (Ger) (1816–92). 1885. Heroult (Fr) and Brothers. Eugene and Alfred Cowles, of Cleveland, Ohio, *inv.* and *const.* first electric furnace (resistance type). 1892. Henri Moissan (Fr) (1852–1907) further *imp.* electric furnace. 1898. Maj. Ernest Stassano, of the Italian army *inv.* arc electric steel furnace. 1916. Induction electric furnace *inv.* by Northrup. 1920. Northrup electric furnace *imp.* by Ribault.

Furniture, Classical English 1754. Thomas Chippendale *des.* furniture. 1788. Hepplewhite *des.* furniture. 1791. Sheraton *des.* furniture.

Fuschine (dye) 1859. First prepared by Verguin.

Fuse (ballistics) 1378. First *inv.* 1421. Earliest known mention of fuse for hollow explosive and incendiary shells. Explosive fuse *inv.* 1831 by William Bickford, of Tuckingmill, Redruth, Cornwall.

Fusible (Safety) Boiler Plug 1802. *Inv.* Richard Trevithick (1771–1833).

Fusee (horology) 1405. According to some authorities fusee was known. 1430. Some

evidence for use of fusee. 1477. *Illus.* and *desc.* by Paulus Alemannus of Rome as being well-known. 1490. *Illus.* and *desc.* by Leonardo da Vinci (1452–1519). 1525. *Inv.* (?) by Jacob Zech (Jacob the Zech, or Lech), of Prague.

G

Gadolinium (element) 1880. *Disc.* Jean Charles Glissard de Marignac (Swit) in rare earths. 1886. Gadolinium named by Marignac in honour of Finnish Chemist J. Gadolin. 1900. First pure gadolinium salts obtained by E. A. Demarcay.

GALEN of Pergamum (130–200) Early biological investigator.

GALILEO, Galilei (It) (1564–1642) *Disc.* fundamental laws relating to the pendulum. First to see lunar mountains and satellites of planet Jupiter. Said to have *inv.* telescope (*q.v.*).

GALL, Franz Joseph of Vienna (1758–1828) 1811. Founded science of craniology or phrenology.

Gallium (element) 1875. *Disc.* by Paul Émile Lecoq de Boisbaudran (Fr).

Gallic Acid 1786. *Disc.* Carl Wilhelm Scheele (1742–86).

GALVANI, Luigi Aloisio (It)

(1737–98) 1762. *Disc.* galvanism (*q.v.*).

Galvanism 1762. *Disc.* Luigi A. Galvani (*see* above). 1791. Galvanism first effectively *desc.* (*See also* Electricity.)

Galvanometer 1820. Original galvanometer *inv.* Johann Salomo Christoph Schweigger (Ger) (1779–1857). 1825. Schweigger galvanometer *imp.* by Leopoldo Nobile (It) (1784–1835), on the astatic needle principle. 1839. Tangent and sine galvanometers *inv.* by Claude-Servain-Mathias Pouillet, watchmaker of Paris. 1867. Mirror galvanometer *inv.* William Thomson, Lord Kelvin (1824–1907). 1903. String galvanometer *inv.* Einthoven (Hol). J. A. Arsonval also *inv.* a mirror galvanometer.

Gamgee Tissue (surgery) *Inv.* Dr. Joseph Sampson Gamgee.

Gangrene 1892. Baccilus *disc.* William Henry Welch (1850–1934) of Baltimore, U.S.

Gas word created by J. B. van Helmont (Bel) (1577–1644).

Gas, Coal 1727. First obtained by distillation of coal by Dr. Stephen Halcs (1677–1761). 1760. George Dixon lit a room in his house by gas at Cockfield, Co. Durham. 1764. French academician M. Jars proposed to light Lyonnais collieries by gas. 1765. Spedding Carlisle, Lord Lonsdale's Whitehaven Colliery manager lit his office by gas. 1667. Gas lighting scheme for Wigan put to the Royal Society. 1760. First fully authenticated attempt to light a room with gas made at Newcastle upon Tyne. 1784. J. P. Minkelers (Bel) proposed to use coal gas for filling balloons; having the previous year originated gas lighting in Belgium with Prof. Philos, of Louvain. 1787. Culross Abbey lit by coal gas by Lord Dundonald. 1792. William Murdock, reputed *inv.* of coal gas lighting, lit his Redruth, Cornwall office by gas. 1807. Phillip Lebon lit Carlton House, London by gas. (Frederick Accum, of Westphalia; Samuel Clegg, of London; and Albert Winsor (Winzler), of Brunswick, often given credit for *intro.* of gas lighting). *c.* 1825. J. G. Appolt *inv.* modern type chamber gas-producing oven, or retort.

Gas, Coal, Production of *c.* 1812. Samuel Clegg (1814–56) *inv.* floating gasometer, gas-holder, gas scrubber, and gas-meter. 1847. Leming *inv.* iron oxide gas purifier. 1855. Beals *inv.* fan gas exhauster for pressurizing mains. 1878. Bunte *inv.* method of analysing gas

and established its calorific value.

Gas Burners 1808. *Inv.* Aimé Argand (Swit) (1755–1803). 1816. Fishtail, or batswing gas burner *intro.* to remain in use substantially unaltered in design until 1900. 1848. Platinum mantle burner *intro.* 1868. J. A. Hogg *intro.* atmospheric gas burner. 1853. R. W. B. Bunsen *inv.* the atmospheric burner bearing his name. 1884. Carl Auer von Welsbach (Ger) (1858–1929) *inv.* incandescent gas-mantle burner; combining the mantle with Bunsen's gas burner in 1890. 1897. Inverted gas-mantle *inv.* by Kent.

Gas-cooker 1802. Albert Winsor (Z. A. Winzler) (Ger) gave a dinner party with food cooked by gas. 1869. Commercial gas-cooker produced and rented to consumers at 6s. per annum.

Gas-discharge Lamp, Electric 1710. Air in glass tube made to glow by passage of electric current by Francis Hauksbee (1687–1763). 1744. Electric gas-discharge lamp proposed for use in German mines. 1857. Becquerel experimentally coated an electric gas-discharge lamp with luminescent material. 1886. Presence of certain impurities in substances used for coating electric gas-discharge lamps known to be necessary. 1904. Lenard and Klatt investigated action of impurities in coating materials. 1923. Rhodamine proposed to improve light from mercury vapour lamps. 1935. Zinc orthosilicate and calcium tungstate used by H. G. Jenkins to *prod.*

yellow light under neon lamp illumination.

Gas-fire 1813. Suggested by John Marban. 1853. First domestic gas-fire *prod.*, to remain unaltered until 1930, when the thermostat was *intro.* in conjunction with it. 1880. Radiant heat gas-fire *intro.* 1860. Gas-ring fire *intro.*

Gas-discharge Lamp *See* Lamp, gas-discharge.

Gas-engine *See* Engine, gas.

Gas "Geyser" (water-heater) 1865. *Inv.* by Benjamin Waddy Maughan.

Gas-lighter 1860. *Inv.* Archibald Wilson, of New York. (Platinum wire and induction coil.)

Gases, Liquefication of 1787. M. van Marum and A. P. Troostwijk (Hol) liquefied air gas. 1823. Michael Faraday (1791–1867) first liquefied chlorine, carbon dioxide, sulphuretted hydrogen, and sulphur dioxide gas. 1835. Carbon dioxide soldified by Thilorier (Fr). 1877. Hydrogen, oxygen, nitrogen, carbon monoxide, air, methane and nitric oxide gases liquefied by Thomas Andrews (1813–85), François-Marie Raoult (Fr) (1830–1901), Marc-Auguste Pictet, of Geneva, and Louis-Paul Cailletet (Fr) (1832–1913). 1897. Air liquefied in large quantities by Karl Paul Gottleib von Linde (Ger) (1842–1934).

Gas-mantle 1885. *Inv.* Carl Auer von Welsbach (Ger) (1858–1929).

Gas-meter 1815. Wet gasmeter *inv.* by Samuel Clegg (1814–56). 1820. Dry (bellows) gas-meter *inv.* John Malam.

Gas, Natural 1000 B.C. First record of use of natural gas at the Oracle of Apollo at Delphi, Greece. 1755. First record of use in U.S., in Virginia. 1821. 30 street lamps lit by natural gas at Fredonia, New York. 1875. Natural gas *disc.* at Netherfield, Sussex, England. 1883. *Disc.* at Medicine Hat, Canada. 1900. *Disc.* at Roma, Queensland, Australia. 1898. Natural gas first used in G.B. for light and power by the London, Brighton and South Coast Railway Co., at their station at Heathfield, Sussex.

Gas-poker 1854. *Intro.* in England.

Gas, "Producer" 1881. First made by J. Emerson Dowson and later *imp.* by (1889) Ludwig Mond (Ger) *inv.* process using cheap slack coal. 1895. B. H. Thwaite used waste gases from blast furnaces to power gas-engines.

Gas-turbine 1791. First *pat.* for taken out by John Barber, of Nuneaton. 1941. First flight with Frank Whittle's gas-turbine. *See also* Aircraft.

Gauges, Engineers' 1856 Bing and plug gauges accurate to 1/1000 in. *inv.* Joseph Whitworth. 1908. Block and slip gauge system *inv.* C. E. Johansson (Swed).

Gauge, Pressure 1847. *Inv.* by Schaffer, of Magdeburg. 1849. Expanding and contracting oval-tube type *inv.* by Edouard Bourdon (Fr) (1808–1884). 1859. Steam cylinder and zigzag, bent spring type *inv.* James Slack, of Nottingham. 1870. Bourdon tube (oval) used by

G. Trouve (Fr) to power model ornithopter.

Gauge, Petrol-tank 1912. Magnetic type *inv.* Thomas Martin. *c.* 1929. Electric type *intro.* on U.S. motor-cars.

GAY-LUSSAC, Joseph-Louis (Fr) (1778–1850) *Inv.* an alcoholometer, a chronometer, and an alkalimeter. *Const.* many balloons and lighter-than-air craft. (All of which *q.v.*)

Gear-wheels and teeth. 330 B.C. Toothed wheels mentioned as being used in windlasses 250 B.C. Gear-wheels used in hydraulic organs and clepshydrae. Two wooden discs with round pegs peripherally set at 90° to each shaft. 200 B.C. Horizontal and vertical shafts coupled by gear-wheels on oxdriven water-lifting machines. 16 B.C. Marcus Vitruvius Pollio (*c.* 50–26 B.C.) *desc.* and *illus.* a pair of gear-wheels for a watermill. (No interest in gears from the fall of Rome until time of Crusaders, who in 13th cent. brought back designs for *const.* of windmills.) 1385. Peter Lightfoot made gear-wheels for clock at Glastonbury Abbey, Somerset. (*See* Clock.) 1660. Machinery near Paris fitted with gear-wheels with cycloidal teeth. 1674. Epicycloidal gearwheel teeth *inv.* Danish astronomer Olaf Roemer (1644–1710); also *att.* to Christiaan Huygens (Dan) (1629–95). 1680. Tooth-shape for crownwheels *inv.* Huygens. 1735. Crown-wheel tooth-shape *imp.* by Charles Étienne Louis Camus (Fr) (1699–1768); and again in 1741 by French mechanic Antoin Thiout. 1750.

De la Hire (Fr) and Camus established geometrical principles of design of gear-wheel teeth Clockmaker John Harrison (1693–1766) used lignum vitae wooden rollers on brass pins as teeth in gear-wheels of his famous chronometers. (*See* Clock.) 1759. John Smeaton *disc.* that cycloidal gear-teeth *prod.* more uniform motion in windmill machinery. 1760. Leonhard Euler (1707–83) *desc.* best shapes for gear-teeth. 1769. Smeaton used cast-iron for windmill and watermill gearteeth. 1781. Sun-and-planet gearing *inv.* by James Watt (1736–1819). 1790. Bevel gearing *inv.* 1790s. Watt used gearwheels of cast-iron with rectangular teeth. 1806–7. Iron spur gears first used in Ireland by Charles Wye Williams. 1813. William Hedley used gearwheels with sloping-sided teeth in his "Puffing Billy" locomotive. 1890. Carl Gustaf Patrik de Laval (Swed) (1845–1913) perfected high-speed helical gearing. (*See also* Gear-cutting machines.) 1923. Novikorshape tooth *prop.* E. Wildhaber (U.S.). 1943. Novikor tooth re-*prop.* by M. L. Novikor (Rus).

Gears, Change-speed 1890. R. Panhard (Fr) *inv.* motor-car change-speed gear with unenclosed gear-"box." 1898. De Dion (Fr) *inv.* motor-car change-speed gears with steering-column control lever. 1901. Louis Renault (Fr) *inv.* motor-car change-speed gear with "tumbler" meshing arrangement. 1902. Premier motor-car fitted with electrically controlled change-speed gears.

1904. Wilson *inv.* his epicyclic pre-selector change-speed gear-box. 1904. Synchromesh change-speed gear *pat.* by Prentice and Shiels. 1904. De Dion *inv.* constant-mesh pre-selector change-speed gear-box. (*See also* Transmission, Variable.) 1905. Linley constant-mesh gear-box.

Gear-cutting Machine 1741. Christopher Polheim Vintro Landop (Swed) *des.* gear-cutting machine. Landop's machine seen by Smeaton and *desc.* as having index-plate and hypoid gear-drive. 1783. Samuel Réhé made gear-cutting machine with milling cutters. 1815. John Buck *inv.* bevel-gear-cutting machine. *c.* 1820. Swiss engineer J. G. Bodmer made gear-cutting machine at his Manchester works. 1835. Joseph Whitworth *pat.* machine for cutting spur and bevel gears. 1837. Prof. Robert *inv.* odontograph for setting out involute teeth. 1840. Josiah Saxton *inv.* gear-generating machine. 1851. First machine-moulded gear-wheels shown at Great Exhibition. 1856. Christian Schiele, of Oldham, England, formulated the germ of the modern gear-hobbing machine. Not known whether one was ever made. 1856. First U.S. gear-hobbing machine, *pat.* 1887. 1867. Clock-maker Potts *pat.* gear dividing and cutting machine and won gold medal at Paris Exhibition of 1867. 1877. Automatic gear-cutters in use in U.S. 1897. Frederick W. Lanchester *inv.* first British gear-hobbing machine. (*See* also Dividing machines.)

6--DOIAD

Gear, Differential 1828. *Inv.* Onésiphore Pecquer (Fr). 1830. John Hanson *pat.* similar gear, and Richard Roberts yet another arrangement in 1832; this being incorporated in a steam road vehicle he built. 1896. Differential gear first applied to motor-cars.

Gear, Variable-speed 1907–12. Hydraulic variable-speed gear *inv.* Harvey D. Williams and Reynold Janney (both U.S.). Included infinitely variable, reversible gear for direct coupling to an electric or other form of high-speed motor.

Gelatine, Blasting 1875. *Inv.* Alfred Nobel (Swed).

Generator, Electric (dynamos and alternators) 1831. Michael Faraday (1791–1867) *prod.* continuous electric current by rotating a copper disc between magnetic poles. 1831. Hippolyte Pixii (Fr) *prod.* alternating current by rotating a permanent magnet near to two fixed coils; and also later the same year added a commutating device to derive direct current from the machine. 1832. William Sturgeon (1783–1850) *prod.* a crude, inefficient electric motor, and in 1835–36 brought the two-part commutator into common use. (*See* Motors, electric.) 1844. J. S. Woolrich *pat.* multipolar, commutator dynamo for electro-plating. 1838. Sturgeon *inv.* shuttle-wound armature; *imp.* by E. Werner von Siemens (1816–92) in 1856. 1858. F. H. Holmes made a magneto-electric dynamo for lighting South Foreland (Kent) lighthouse. 1860. Antonio Pacinotti (It) (1841–

1912) *prod.* a ring-wound armature. 1863. Henry Wild *inv.* his separately excited dynamo. 1866–67. E. Werner von Siemens and S. A. Varley (1832–1921) working with Sir Charles Wheatstone (1802–75) *dev.* a practical self-exciting dynamo which relied on its residual magnetic field. 1870. Zénobe Théophile Gramme *intro.* practical ring armature which was *dev.* by F. von Hefner-Alteneck and used by Siemens Brothers in 1873. 1887. Nikola Tesla (1857–1943) *pat.* two-phase alternators and induction motors (*q.v.*) thus inaugurating polyphase electrical engineering. *See also* Motors, electric, and Transmission, electric.

Generator, Thermo-magnetic Electric 1887. Thomas Alva Edison (U.S.) (1847–1931) *pat.* idea of thermomagnetic electric generator, using magnetic effects of heat. 1960. J. F. Elliot (U.S.) suggested use of gadolinium for thermo-magnetic electric generators.

Geography, Science of 443 B.C. Established by Herodotus. 240 B.C. Systematized by Eratosthenes. 150 B.C. Expounded by Ptolemy. 24 B.C. Further *dev.* by Strabo.

Geology 1779. Word *intro.* by H. B. de Saussure, of Geneva.

Geological Formations, Names of Devonian, Carboniferous, Pliocene, and Miocene named by Sir Charles Lyell (1797–1875); Cambrian, Palæzoic and Cainzoic by Sedgwick, of Cambridge; and the Mesozoic by John Phillips, of Oxford.

Geometry 600 B.C. *Intro.* into Greece by Phales. Geometry signs *intro.* by Chasles (1793–1880). Co-ordinate geometry *prop.* by Réne Descartes (Fr) (1596–1650) and Pierre Fermat (Fr) (1601–65). 1768. Descriptive geometry conceived by Gaspard Monge (Fr) (1746–1818).

Germanium (element) 1886. *Disc.* by Clemens Winkler (1838–).

Gettering Process 1882. Magnesium process *inv.* Fitzgerald. 1894. Phosphorus process *inv.* Arturo Malignani (It).

Ghost", "Pepper's 1858. Optical illusion first *prod.* by John Taylor. 1863. Prof. Pepper exhibited illusion at Royal Polytechnic, London.

GIABER (Geber, or **Yeber)** (Arab) (*c.* A.D. 800) *Disc.* nitric acid, nitro-hydrochloric acid, silver nitrate, gold chloride, sulphuric acid, mercuric chloride (corrosive sublimate), and mercuric oxide (red oxide of mercury).

GIFFARD, Henri (Fr) (1825–82) *Inv.* the injector for steam boilers; a 3 h.p. steam-driven dirigible airship in 1852; and *des.* and *const.* a 230 ft long airship in 1855.

Gilding 149 B.C. Gilding employed as an art by Moses (Exodus xxv. 11). 1273. Gilding with gold-leaf *inv.* in Italy.

Gimbal Suspension 2nd cent. B.C. *Desc.* by Philon of Byzanteum. 1st cent. B.C. Known to the Chinese. (A.D.). 12th cent. mentioned in Western Civilization. 1501 *Att.* to Jerome Cardanus (It). 1546. First *desc.* by Martin Cortez.

Glacier-flow 1840-42. Speed

of *disc.* by Louis Agassiz (Fr) and J. D. Forbes.

Glanders Bacillus *Disc.* by Friedrich Loeffler (Ger) (1852–1925).

Glass 1st cent. B.C. Glazing of windows a general practice in Rome. (Bronze window-frames glazed with sheets of glass 28 in. by 21 in. found in ruins of Pompeii.) Blue glass cylinder of *c.* 2600 B.C. found at Tel Asmar, north-west of Baghdad. *c.* 15th–16th cent. B.C. Egyptian glass-works found by Dr. Flinders Petrie at Tel-el-Amarna in 1894. Moulded and carved glass objects, 7th cent. B.C. found at Puzzuoli; of 4th cent. B.C. at Ephesus and Artemesium; of 3rd cent. B.C. at Canosa; and of 2nd cent. B.C. at Aegina. 3rd–2nd cent. B.C. "Millifiore" glass in *prod.* 1st cent. B.C. Earliest blown glass objects made in Egypt by blowing glass into moulds. (By A.D. 4th cent. blown glass was *prod.* in Italy, Greece, Asia Minor, Kertch, South Russia, Melos, Tharros, Sardinia, Genoa, Smyrna, Crete, Cyprus and Palestine. A.D. 220. Tax imposed on glassworkers of Ancient Rome. 674. Venerable Bede mentions Gaulish glass-workers imported by Abbot Benedict to make glass for windows of Wearmouth, Northumberland, church. 5th–9th cent. Coloured glass used for decorating church windows. (Anastasius, librarian to Pope Leo III (*c.* 800), mentions painted glass windows as in use in his time; and Leo Ostiensis (*c.* 760) refers to glass windows fixed with lead and iron rods.)

1023 Manuscript encyclopaedia alludes to free-blown glasswork with *illus.* of blowing and annealing. 12th cent. German monk Theophilus of Paderborn, Westphalia, *desc.* production and working of glass together with *const.* of glass furnace. 1380. John Glasewrythe, of Kirdford, Sussex, made "brode-glass" (window-glass) and hollow-ware vessels. 1386. Charter of King Richard II mentions glass window manufacture. 1567. Lorranian Jean Carré obtained licence to make glass at Alford, Surrey and Crutched Friars, London. 1572. Venetian Jacomo Verzalini made Venice glass in London. 1610. Thomas Percivall *pat.* use of coal as fuel for glass-making. 1633. Thomas Tilman *prod.* crystal-lead glass. 1664. Glass Sellers Company chartered by King Charles II. 1665. New casting method *disc.* by Nehou. 1675. George Ravenscroft *inv.* flint glass to rival Venetian crystal glass. 1675. Crystal glass (the "chalk-glass" of Bohemia) appeared. 1688. Plate glass *intro.* in France. 1688. Nehou's casting method *imp.* by Louis Lucas (Fr). 1894. Michael J. Owens (U.S.) *inv.* semi-automatic paste-mould electric-lamp-bulb blowing machine. *See also* Bottle-making machine, Automatic.

Glass, Boron Originally *inv.* Michael Faraday (1791–1867) and brought into commercial use after World War II.

Glass Building-blocks. 1931. First. *intro.*

Glass, Fibre 1841. Louis

Schwabe demonstrated glass fibre spinning at Manchester. 1930. Staple glass fibre first *intro.* 1944. Glass fibres first used to reinforce certain contact-pressure resins.

Glass, Optical 1733. Chester Moor Hall used flint and crown glass lenses to counter chromatism. (*Also* John Dolland, in 1758.) 1755. Flint and crown glass used for lenses by Samuel Klingenstierna (Swed). (Optical glass industry founded by Pierre-Louis Guinand (Swit) (1774–1802).

Glass, Plate 1687, first mechanized in U.S. 1688. *Intro.* into England. 1773. *Intro.* into France. 1900–12. Colburn *dev.* machine for making plate glass by continuous process. 1910. Bernard Perrot's process, *inv.* 1920. Perrot's mechanized process made continuous at Ford's Motor Works, U.S. 1952. Alastair Pilkington of St. Helens, Lancashire *inv.* "float" process for making plate glass.

Glass, Safety 1905. Wood *inv.* laminated (celluloid) safety glass. 1910. E. Benedictus *prod.* first commercial safety glass ("Triplex"). 1929. Safety glass *prod.* by St. Gobian Glass Company, France.

Glass, Toughened 1785. *Inv.* by François de la Bastie, of Paris. 1876. Use of toughened glass general in France.

Glasses, Musical (harmonica) 1641. *Inv.* by Kauffmann, of Nuremberg. 1746. First played in public in London by musical composer Christopher Willibald von Gluck (1716–1787). (According to some authorities musical glasses

inv. 1697 by Richard Pockridge.)

Glauber Salts (sodium sulphate) 1658. First made by Dr. Johann Rudolph Glauber, of Bavaria (1604–70).

Glazing (ceramics) 1283. Lucca della Robbia *inv.* tinglazing of pottery. 1671. Dr. Dwight *pat.* white salt-glazed pottery and stoneware.

G.L.E.E.P. (graphite low energy experimental (atomic pile)) 1947. *Des.* and built at Harwell, England.

Globes, Terrestial and Celestial 4th cent. B.C. Globes 6 ft. 6 in. in diameter made for astronomer Eudoxus (*d.* 386 B.C.). (Now in Naples Museum.) Archimedes and Heron of Alexandria had glass celestial globes with solid terrestial globes inside them.

Glockenspiel (Mustel organ) 1886. *Inv.* by Auguste Mustel, of Paris.

Gloves 1225. Gauntlets *intro.* into England. 1889. Rubber gloves first used in surgery by William Stewart Halstead, of Baltimore, U.S.

Glue 1600 B.C. Glue of similar composition to that used today in use in Ancient Egypt.

Glycerine (glycerol) 1779. *Disc.* by Carl Wilhelm Scheele (1742–86) and termed by him "oelsüss."

Glycogen (chemistry) *Disc.* Claude Bernard (1813–78), of St. Julien, Lyons, France.

Goitre, Exopthalmic (surgery) First identified by Robert Graves (1796–1853).

Gold Pre 5000 B.C. *Disc. c.* 1350. Gold wire first made in Italy. *Disc.* in South America by

Spaniards, 1492; in Malacca, 1731; in Ceylon, 1800; in California, 1847; in Australia, 1851; in British Columbia, 1858; in New Zealand and Nova Scotia, 1861. 1867. Miller *inv.* chlorine process of separating gold from silver. 1889. McArthur-Forrest *inv.* cyanide extraction process.

Golden Number (19-year cycle) 432 B.C. *Inv.* by Meton of Athens.

Golgi Bodies (cytology) 1909. Investigated and named by Cammilo Golgi (1884–1906), of Pavia, Italy.

GOMPERTZ, Lewis *Inv.* three-point, self-centring lathe chuck.

Goniometer, Reflecting *Inv.* Étienne-Louis Malus (Fr) (1775–1812). *c.* 1857. Simple goniometer *inv.* Woolaston.

Governor, Centrifugal *c.* 1490. Francesco di Giorgio depicted ball-and-chain centrifugal governor as Tibetan hand prayer-wheel in conjunction with compound cranks and connecting-rods. 1507. Rotation of roasting-spit regularized by three weights turning on vertical axis. Centrifugal governor *inv.* by James Watt (1736–1819) and applied to his steam-engine.

Graafian Follicles (anatomy) First *disc.* by Regnier de Graaf (1641–73).

Graduation (mechanical division 1766. Jesse Ramsden (1735–1800) completed his dividing engine. 1778. John Troughton made a dividing engine. 1830. Andrew Ross *inv.* a new dividing engine. 1843. Sims applied self-acting principle to Edward Troughton's circular dividing engine. (Robert Hooke (1635–1703) and Ole Rømer (1644–1710) *intro.* earliest methods of mechanical division.)

Grain-cleaning Machines 1677. Edward Melthorpe and Charles Milson *inv.* machine for hulling barley and pepper. 1715. Thomas Martin *inv.* machine for cleaning maize. 1725. George Woodroffe *inv.* water-driven machine for cleaning wheat; and Robert Burlow a similar machine in 1731. *See also* Winnowing Machine.

GRAMME, Zénobe Théophile (Ger) (1826–1901) Responsible for many *invs.* in the field of electrical power production and transmission. (*See* Electrical generators, alternators and motors.)

Gramophone 1887. Emil Berliner (Ger) *inv.* disc record gramophone. The method of duplicating records by matrix was *inv.* by Bettini (It) at a later date. (*See also* Phonograph.)

Graphotyping 1860. Method of printing *inv.* Hitchcock, of New York.

Grating, Diffraction (optics) First made by Joseph von Fraunhofer (1787–1826).

Gravitation, Theory of 1665. *Prop.* by Sir Isaac Newton (1642–1727). 1679. Newton's theory completely proved. 1743. Gravitational theory bearing his name put forward by Alexis-Claude Clairaut (Fr) (1713–65).

GRAY, Stephen (1670–1736) 1729. First distinguished between conductors and non-conductors of electricity.

GREW, Nehemiah (1641–1712) Botanist who *disc.* flowers to be sex-organs of plants.

Grimm's Law (grammar) Law of transmutation of consonents *disc.* by Wilhelm Carl Grimm (Ger) (1786–1859).

Grinder, Surface (Leonardo da Vinci (1452–1519) sketched a disc grinder and internal and external grinders and polishers.) 1831. J. W. Stone (U.S.) *pat.* horizontal table, with power-driven rack or screw traverser. 1845. James Nasmyth (1808–90) *inv.* disc grinder with two annular wheels 7 in. diameter set with 12 stones in segments. 1853. Samuel Darling *inv.* a surface grinder.

Grinding, Centreless 1915. *Intro.* by L. R. Heim (U.S.).

Grindstone 816–834. Earliest rotary grindstone depicted in Utrecht Psalter. 1340. Double-cranked grindstone *illus.* in Lutrell Psalter.

Grooving and Tongueing Machines 1793. *Inv.* Sir Samuel Bentham. 1827. *Inv.* by an employee of Henry Maudslay (1711–1832). 1860. Groov-and slot-drilling machine with flat, arrow-head, centreless bit *inv.* James Nasmyth.

GROSSETESTE, Robert (*c.* 1175–1253) First Chancellor of Oxford University. *Prop.* mathematical basis of science. Devised the hypothetico-deductive method of scientific reasoning. Student of the metaphysics of light.

GROVE, Sir William Robert (1811–96) *Inv.* electric battery bearing his name, and a gas battery for decomposition of water and potassium iodide.

GUERICKE, Otto von (Ger) (1602–86) *Inv.* hygroscope, valveless vacuum air-pump.

Guillotine 1785. *Inv.* Joseph Ignace Guillotin (Fr) (1785–1814). 1792. Guillotine first used.

Gun See Cannon.

Gun-carriage 1763. *Inv.* Claude François Berthollet (1718–1800).

Guncotton 1846. *Disc.* Schönbein, of Basle, and by Johann Friedrich Bottger (1685–1719), independently. *Pat.* 1847.

Gunpowder 200 B.C. Used by Chinese in rockets. 673 Syrian refugee architect Kallinikos *inv.* "Greek fire." 690. Used by Arabs at siege of Mecca. 850. Used as an explosive in China, where recipe published in 1040. 1118. Used by Arabs in cannon —Moors against Saragossa. 1216. Earliest record of composition of gunpowder in the west in Roger Bacon's *Secrets of Art and Nature*. 1231. Gunpowder grenades used by Chinese. 1306. Used by Moors during siege of Gibraltar. 1325. Used against the King of Granada at siege of Baza. 1429. Gunpowder given qualities of more rapid burning and greater consistency of action by corning. 1613. First used for mine blasting.

Gunter's Scale (land surveying) 1620. *Inv.* Edmund Gunter, London. In *c.* 1600 he had *inv.* the 100-link chain.

GUTENBERG, Johann Gensfleisch (Ger) (*c.* 1397–1468) *b.* Mentz. Pioneered the art of printing (*q.v.*) in Germany.

Gutta Percha *Intro.* into Europe

1822. 1849. First used as electric cable insulator by Michael Faraday (1791–1867). (*See also* Rubber.)

Guyot (underwater "island") *c.* 1940. First *disc.* by geologist Henry Hess, of Princeton, U.S., and named after Arnold Guyot, prof. of geology at Princeton University.

Gyro Compass Léon Foucault (Fr) (1819–68) showed that a gyroscope (*q.v.*) would point north. 1865. Trouvé (Fr) made an electrically driven gyroscope. 1878. Hopkins *inv.* electrically driven gyroscope. 1903. Dr. Herman Anschütz-Kaempfe (Ger) *pat.* the first gyro-compass. 1908. Elmer Ambrose Sperry (U.S.) *pat.* gyro-compass. 1914–18 S. G. Brown *inv.* gyro-compass, in collaboration with Prof. John Perry. *See below.*

Gyroscope 1816–17 Early type of gyroscope already in existence. 1836. Edward Sang (Scot) suggested that gyroscope could prove rotation of the earth, but no one carried out the experiment until 1852, when Foucault did so and named the gyroscope. (The gyroscope said to have been used by Serson for steering a ship, which later foundered.) *See above.*

Gyro Stabilizer 1868. Germ of gyro stabilizer idea ("to keep a craft on a fixed vertical and horizontal course") put forward by Matthew Piers Watt-Boulton. 1903. Schlick (Ger) first applied gyro stabilizer to ship. 1913. Elmer Sperry (U.S.) began fitting gyro stabilizers to U.S. warships and merchant vessels. (*See also* Automatic pilot.)

Hackney Carriage 1625. Originated in London, where first stand was established in the Strand in 1634 by Capt. Baily.

Hafnium (element) 1845. Found, but unidentified by Svenborg. 1911. Re-*disc.* 1923. Finally *isol.*

HAHN, O. Atom scientist who, in 1938, first gave chemical evidence that uranium, when bombarded by neutrons, produced atoms of lower mass—notably barium.

Hair, False 1572. First *intro.* into England from France.

HALES, Stephen (1677–1761) First *disc.* that plants absorbed air.

HALLEY, Edmund (1656–1742) 1682 *Disc.* periodic comet now bearing his name.

Hammer, Explosion 1870. *Inv.* Phillip Sing Justice, of Pennsylvania, U.S.

Hammer, Steam 1784. According to Thurston, steam hammer *inv.* by James Watt (1736–1819). 1806. William Deverill *pat.* steam hammer. 1838. *Inv.* by James Nasmyth (1808–90). 1850. Double-acting steam hammer *inv.* 1860. Thomas Fearnley *pat.* method of governing length of stroke.

Hammer, Trip 4th cent. B.C. In use in China.

Handkerchief 1743. First one in G.B. made at Paisley.

Hansom Cab 1833. Safety cab *pat.* and *intro.* by J. A. Hansom.

Hardening, Metal 1906. Wilm accidentally *disc.* age-hardening property of aluminium, which led to *inv.* of duralumin (*q.v.*).

HARGREAVES, James (1720–1778) 1767. *Inv.* the spinning "jenny."

Harmonic Curve 1714. First demonstrated by Dr. Brook Taylor. Later perfected by D'Alembert, Euler, Bernouilli and Lagrange.

Harmonichord (Keyed instrument) 1810. *Inv.* by Kauffmann (Ger).

Harmonicon ("Singing Flame") *Inv.* John Tyndall (1820–93).

Harmonics 1754. *Disc.* by violinist Guiseppe Tartini (It) (1692–1770).

Harmonium 1810. *Inv.* by Grenie, of Paris. 1840. *Imp.* by Alexandre Debain (Fr). 1841. *Imp.* by Evans, of Cheltenham, Glos.

Harmony, Science of 1482. Equal interval scale devised by Bartolo Rames (Sp). 1511. Equalized scale *prop.* by organ-builder Arnolt Schlick (Ger). 1577. Schlick's theory elaborated by Franciscus Salinus.

Harness, Horse *c.* 1487 B.C. Said to have been *inv.* by Erichthonius of Athens. A.D. 800. Horse harness with collar *illus.* in manuscript at Triers (France) city library.

Harp 3875 B.C. Mentioned in the Bible—Genesis iv. 21. *c.* A.D. 1012. Brian Boromu harp (taken there in 1782). 1720. Hochrücker, of Donauwörth, Bavaria *prod.* hand-operated harp. 1810. Hand and foot operated harp (double-action) *inv.* Sebastian Erard; with more *imps. pat.* 1895.

Harpoon-gun 1731. *Inv.* 1772. *Imp.* by Abraham Staghold.

Harpsichord 1409. First mentioned, as "Grand harpsichord." 1512. First made in Germany. 1522. *Intro.* into Italy. 1667. *Intro.* into England. 1708. New form *inv.* by Cursinie. 1716. Jack-and-quill action replaced by hammers by Marius. 1754. Ocular harpsichord *inv.* by Jesuit Father Louis Bertrand Castel, of Montpelier. *See also* Pianoforte.

HARRISON, Dr. John (of Australia) In 1857 installed first refrigeration equipment (sulphuric ether). (*See* Refrigerator.)

HARRISON, John (1693–1776) of Barrow-upon-Humber. Famed for his accurate chronometers. *Inv.* gridiron temperature-compensating clock pendulum.

Harrow, Disc 1868. *Inv.*

HARVEY, William (1578–

1657) In 1615 *disc*. circulation of blood in the human body. (*Disc.* not announced until 1628.)

HAUTFEUILLE, Abbé de (Fr) 1678. *Prop*. engine working by gunpowder. (*See* Engine, internal-combustion.)

Heart, Contraction of the (and Pulse) *c*. 300 B.C. First description of, together with speed and strength of beat.

Heat 1760. Latent heat *disc*. by Joseph Black (1728–89). 1819. P. L. Dulong (Fr) and A. T. Petit *pub*. paper linking atomic theory with that of heat. 1822. Theory of heat *prop*. by J. B. J. Fournier (Fr). 1824. Theory of heat propounded by Sadi Carnot (Fr) (1796–1832). 1843. Kinetic theory of heat *prop*. by James Prescott Joule (1818–89), of Manchester.

Heating, Atomic 1951. First B.E.P.O. plant opened at Harwell, England.

Heating of Buildings, etc. 95 B.C. Central heating *inv*. by Sergius Orata for heating fish and oyster ponds at Baiae, on the shore of Lake Lucrine, Italy. 1716. Hot-water pipes for greenhouses *inv*. by Sir William Triewald (Swed). 1800. Messrs. Todd and Stevenson's mills steam-heated. 1814. William Murdock heated his conservatory by steam, and the following year heated Leamington Spa baths by same method. 1830. Ernst Alban steam-heated his factory at Plau, Mecklenburg.

Heat control 18th cent. Lhomond (Fr) *inv*. shutter-type heat control damper. *See also* Thermostat.

Heat-pump (reversed refrigeration cycle) 1852. Suggested by William Thomson (Lord Kelvin) (1824–1907). 1945. First building (municipal building at Norwich, Norfolk) so heated.

HECKEL, Wilhelm (Ger) *Inv*. modern contra-bassoon.

Helicopter (*see also* Autogiro) A.D. 320. Used as toy top in China; and in Western Europe in 1200. 1505. Leonardo da Vinci (It) (1452–1419) *desc*. helicopter in book *Sul volvo de gli Uccelli*. Book not known to world until 1797. 1768. Paucton (Fr) *inv*. "Pterophore," with lifting and propelling airscrews. 1784. Launoy and Bienvenu (Fr) made successful model with 21 ft. diameter feathered, contra-rotating screws. Driven by bow and twisted cord. First modern helicopter. 1828. Florentine cobbler Vittorio Sarti *des*. helicopter with four-bladed, contra-rotational lifting screw. 1842. W. H. Phillips made steam model helicopter with steam fed through rotor-blades to jets at tips. First power-driven model helicopter to fly. 1861. Viscomte de Ponton d'Amécourt (Fr) *pat*. contra-rotational air-screw (steam-driven). 1863. Helicopter design a rage among European inventors. 1886. Jules Verne *des*. fictional helicopter, the "Albatross," 125 m.p.h., with a ceiling of 8,700 ft. (Not until 1962 was this fictional performance realized in practice, when Igor Sikorsky's "S-62" reached 110 m.p.h. with a ceiling of 8,500 ft.) 1904. Charles Renard (Fr) *inv*. articulation of rotor-blades to their hub. 1906. G. A. Rocco (It) *inv.* cyclic pitch control. 1907. First man-carrying helicopter flights made by Brothers Louis and Cornu Breguet (Fr). 1912. Cyclic pitch blade first used by Ellehammer (Dan). 1919–25 Marquis de Pescara (Argentine) first obtained horizontal propulsion from lifting rotor. 1922. Autogiro *inv.* Juan de la Cierva (Sp). 1934. Dr. Henrich Focke (Ger) *const.* helicopter, which first flew successfully 1938, with twin rotors (known as the Focke-Achgelis F.W.61. Igor Sikorsky (Rus) flew the machine). *See also* Aircraft.

Heliometer 1748. *Inv.* by Pierre Bouger, of Brittany. 1754. *Imp.* by Joseph Jérôme de Lalande (Fr) (1732–1807).

Helioscope 1625. *Inv.* Christopher Schiener (Ger).

Helium (gas) 1867. *Disc.* spectroscopically in spectrum of sun by Sir Norman Lockyer. 1882. *Disc.* on earth by Prof. Palmieri (It). 1905. Found in natural gas in U.S. by Messrs. Cady and McFarlane. 1909. Helium first liquefied by Kamerlingh Onnes, of Leyden, Holland. 1943. World's largest helium plant opened at Excel, Texas, U.S.

HELMONT, Johann Baptiste van (Bel) (1577–1644) Studied pneumatic chemistry and gases, creating the term "gas."

Hemisphere (scientific instrument) 330 B.C. *Inv.* by Babylonian astronomer-priest, of Cos.

HENSON, William Samuel (1805–88) of Chard, Somerset.

Pat. "aerial steam carriage," lace-making machinery, and the safety-razor. Also interested in the problem of flight. (*See also* John Stringfellow.)

HERSCHEL, Wilhelm Friedrich (Ger) (1738–1822) In 1800 *disc.* infra-red rays.

HERSCHEL, Sir John Frederick William (1792–1871) In 1845 *disc.* luminescence of quinine sulphate. Continued his father's astronomical investigations.

HERTZ, Rudolf Heinrich (Ger) (1857–94) First *prod.* radio waves experimentally.

Hieroglyphics c. 1830. Decyphered by Jean François Champollion (Fr) (1790–1832), following a hint from Dr. Thomas Young (1773–1829). (*See also* Cuneiform.)

HILL, Sir Rowland 1840 (Jan. 10). British penny postage instituted. *Inv.* by him.

Hinge 1780. Rising door-hinge *inv.* by Gascoigne.

Histology Science founded by Antonie van Leeuwenhoek (Hol) (1632–1723). 1819. Word *intro.* by A. F. J. K. Meyer. 1852. Histology first defined by Quackett.

Hob, Gear-cutting 1856. Suggested by Christian Schiele. 1887. G. B. Grant (U.S.) made first hobbing machine for spur gears. 1896. Frederic W. Lanchester *const.* first hobbing machine for cutting worm gears. 1896. E. R. Fellows *inv.* entirely new gear-shaping machine. 1897. Herman Pfanter (Ger) *inv.* machine for cutting spiral and spur gears. (*See also* Gears.)

Hoe, Horse 1714. *Intro.* from France by Jethro Tull.

HOHENHEIM, Phillipus Aurelius Theophrastus Bombastus von (Paracelsus) (Swit) (1493–1541) Made first attempt to equate chemical action with bodily processes. (*See also* van Helmont.)

Hoist, Steam 1830. First *dev.*

Hollander (paper-making) Late 17th cent. *Inv.* by a Dutchman, hence its name. 1712. First hollander installed in Germany.

Holmium (element) 1878. *Disc.* by J. L. Soret (Swit). 1879. Independently *disc.* by P. T. Cleve (Swed).

Homeopathy 1796. Founded by C. F. S. Hahnemann (Ger) (1775–1843). 1827. *Intro.* into England. 1845. Homeopathy Association founded. 1859. Homeopathy Society founded. 1876. London School of Homeopathy founded.

Honey 5510 B.C. First record of bees found in tombs at Abydos, Egypt. 2000 B.C. Honey taken by King Kamose in expedition against the Shepherd Kings (Hyksos). Wax also mentioned. A.D. 3rd cent. First European mention of honey in Brehon Laws of Ireland. Domesday Book lists all beehives in Britain.

HOOKE, Dr. Robert (1635–1703) "Hooke's Law" *disc.* 1660. 1667. *Inv.* drawn-wire "otacousticon," or telephone. 1680. *Inv.* screw propeller (*q.v.*). 1684. *Inv.* first practical semaphore telegraph.

HOPPUS, Edward 1846. *Inv.* and *pat.* measuring tables showing solid content of timber, etc.

Hops 1425. First mentioned in England.

Horizon, Artificial 1929. *Inv.* A. H. Sperry.

HORNBLOWER, Jabez Carter (1753–1815) 1781. *Des.* and *pat.* first successful compound steam-engine, installed at Radstock, Somerset, in 1782.

Horse-power Unit *intro.* by James Watt (1736–1819).

Horseshoe 4th–5th cents. Nailed horseshoes of this date found by Sir Mortimer Wheeler at Maiden Newton, Dorsetshire. 9th–10th cents. Horseshoes found of this date in Yenesei Region of Siberia. 886–911. Mentioned by Leo VI of Byzantium. 973. Nailed horseshoes mentioned by Gerard in *Miracula Sancti Oudalrici*. 1038. Silver-nailed horseshoes used by Boniface of Tuscany. 1066. Six blacksmiths worked at Hereford. 1860. Goodenough (U.S.) *intro.* machine for making horseshoes.

"Hot-spot" (automobile engine) 1904. First internal combustion engine to be fitted with hot-water gas inlet muffle a De Dion.

Hour-glass 1306–13. First mentioned (as "arlogio") by Francisco de Berberino. 1441. First pictured by Petrus Christos. 15th cent. (late). First represented as attribute of Father Time.

Hovercraft (air-cushion vehicle) 1958. British boatbuilder Christopher Cockerell demonstrated principle. 1958. Dr. Andrew A. Kucher proposed and built a 36 ft. long "Levacar" (Ford). 1959. Curtis-Wright (U.S.) *prod.*

"2500" "G.E.M." (ground effect machine) which had 45 h.p. engine and travelled at 75 m.p.h. 1959. Saunders-Roe *prod.*, in Isle of Wight, SNR-I hovercraft, which later same year made the first open-sea voyage from Calais to Dover.

HUMBOLDT, Baron Alexander von (1769–1859) Distinguished German savant. Studied composition of the atmosphere with Joseph Gay-Lussac (1778–1850).

HUYGENS, Christiaan (1629–95), of Zulichem, Holland. *Disc.* planet Saturn's rings and largest satellite. *Prop.* undulatory theory of light. Advanced first theory of motion of pendulum. Made first marine timekeeper (of which a replica still works in Amsterdam Museum). 1690. Recognized and explained polarized light.

"Hydramatic" transmission Coupling (automobile) 1930 (early). Semi-automatic hydramatic transmission coupling *inv.* Earl A. Thompson (U.S.). 1930s. A. W. Hallpike and A. A. Miller (G.B.) *inv.* auto-controlled epicyclic gearboxes. (*See* Gear-box and Preselector gears.) 1939. Earl A. Thompson *dev.* fully automatic hydramatic transmission coupling.

Hydraulic Transmission Coupling (fluid flywheel) 1926. First applied to vehicle by Harold Sinclair (U.S.). 1930. Used by Daimler Motors in conjunction with Wilson (E.N.V.) pre-selector gear-box (*q.v.*). 1935 (by) Hydraulic

transmission coupling *const.* to transmit up to 36,000 h.p.

Hydrazine (chemical) 1875. *Disc.* by Emil Fischer (Ger) (1852–1919).

Hydrochloric Acid 15th cent. *Prod.* from salt and alum.

Hydrofluoric Acid 1780. *Disc.* Carl Wilhelm Scheele (Swed) (1742–86).

Hydrogen 1766. *Disc.* Henry Cavendish (1731–1810). 1781. Cavendish, with James Watt proved water to be the sole product when hydrogen was burnt.

Hydrokineter (boiler-water circulator) 1874. *Inv.* James Weir, of Glasgow.

Hydrometer 400. *Inv.* by Hypatia of Alexandria. *c.* 1830. Bead-type hydrometer *inv.* by Lovi. *c.* 1830. Sykes hydrometer. *inv*

Hydropathy 1826. *Inv.* by Vicenz Priessnitz.

Hydrophobia *See* Virus.

Hygrometer *Inv.* by Sir John Leslie (1766–1832).

Hygroscope *Inv.* Otto von Guericke (1602–86). 1665. *Inv.* Robert Hooke (1635–1703). 1687. Another type *inv.* by Guillaume Amontons (Fr) (1663–1705). 1751. By Dulac (Fr). 1772. By Johann Heinrich Lambert (Ger) (1728–77). 1783. Hair hygroscope *inv.* by Horace-Benedict De Saussure (Fr) (1740–99).

Hypochlorites (and **Chlorites**) *Disc.* C. L. Berthollet (Fr) (1748–1822).

Hypsometer *Inv.* William Hyde Woollaston (1766–1828).

I

Ice 1786. First *prod.* by mechanical means.

Ice-cream 1924. First made by CO_2 process in New York.

Ichnology (footprints) 1828. Science of founded by Dr. Duncan.

Iconoscope *See* Cathode-ray tube.

Icthyology 1553–58. Science of founded in France.

Ignition, Motor-car Engine *c.* 1890. Robert Bosch *des.* and made low-tension magneto for gas-engines. 1897. Bosch low-tension magneto used on motor-cars—on first Daimler-Mercedes car in 1901. 1902. Gottlob Honold, of Bosch's firm *inv.* first high-tension magneto, and first Bosch sparking-plugs made at Bamberg, Bavaria. 1903. Capt. Longridge *inv.* automatic advance-retard mechanism for magnetos. *See also* Transformer, Electrical and Coil, induction.

Impact Testing Machine *c.* 1919. *Inv.* Charpy.

Incubator 1588 Giovanni Battista della Porta (1535–1615) *des.* an incubator after plan of those used in Ancient Egypt. 1666. Cornelius Drebbel (Hol), of Alkmaar *inv.* incubator ("Athenor") fitted with a thermostat). 1770. John Champion, of London, *pat.* incubator. 1846. Re-*pat.*

Indicator, Internal-combustion Engine 1923. "Farn-

borough" indicator *intro.* to check performance of aero engines in flight. 1940. Electronically operated indicators first appeared.

Indicator, Steam-engine 1794. *Inv.* James Watt (1736–1819).

Indigo 1516. Plant first known in England. 1600. First known in Germany. 1747. First planted in Carolina, U.S. 1840. First imported into England.

Indigo, Synthetic 1856. Synthesized from benzol. 1880. Synthesized from nitro-cinammic acid by Adolf von Beyer and Emerling (Ger).

Indium (element) *Disc.* by Jeremias Benjamin Richter in zincblende.

Induction-coil *See* Transformer, High-tension.

Influenza Bacillus *disc.* Richard Pfeiffer (1909–26), of Breslau.

Infra-red Rays 1800. *Disc.* W. F. Herschel (Ger). 1847. Re-*disc.* J. W. Draper (1811–82).

Inhaler, Ether (medical) *Inv.* John Snow (1813–58).

Injector, Steam boiler 1753. Capt. Savery *pub.* plan and *desc.* of conical jet steam nozzle for lifting water. 1858. *Inv.* Henri Giffard (Fr) (1825–82). 1860. Manufactured in U.S. and *appl.* to locomotive engines by Matthias Baldwin. Pre-1861. *Imp.* by August Nagel, of Hamburg.

Ink, Indian 800. First made in China.

Ink, Invisible 1653. *Inv.* Peter Borel. 1684. Another *inv.* by Le Mort. 1705. Yet another type by Waitz. 1737. Cobalt type *inv.* Hellot (Fr). (Ovid suggested writing with fresh milk and warming writing to discover it.)

Ink, Modern Writing 1836. Henry Stevens *inv.* writing fluid and a fountain-type ink-pot. 1849. *Inv.* Runge.

Inlaying, Metal 1600 B.C. Craft practised on ivory and wood in Ancient Egypt. (Bronze electrum daggers inlayed with silver and gold in lion-hunt scene.)

Inoculation 1721. *Intro.* into England from Turkey, and tried on condemned criminals. 1723. *Intro.* into Ireland. 1724. Into Hanover. 1726. Into Scotland. 1796. Vaccine lymph *disc.* Dr. Jenner. Inoculation against small-pox *disc.* Baron Dimsdale (1711–1800). Immunization by inoculation *disc.* Robert Koch (Ger) (1843–1910).

Insecticide 1924. First recorded use of. (*See also* D.D.T.)

Insulin 1921. *Disc.* by Frederick Grant Banting (Canada) (1891–1941) and Charles Hubert Best (1899–). 1922 (Jan. 11). First human patients received insulin treatment at Toronto General Hospital. 1926. *Isol.* by J. J. Abel. 1934. D. A. Scott *prod.* zinc salt of insulin.

Integraph 1878. *Inv.* Abdank Abakanovicz. 1882. *Inv.* or *imp.* by C. Vernon Boys (1852–1900).

Interferometer 1892. First interferometer *inv.* A. A. Michelson. *Imp.*, 1905 by C. Faby and A. Perot (Fr).

Invar (Nickel-iron alloy) 1896. *Disc.* Dr. Charles E. Guillaume, of International Bureau of Weights and Measures, Paris.

Iodoform 1822. *Disc.* by Georges S. Serrulas (1744–1832). 1878. Iodoform first used as an antiseptic.

Iodine 1812. Iodine *disc.* by Parisian saltpetre manufacturer Bernard de Courtois. 1813. Hydrogen Iodide jointly *disc.* by MM. Desormes and Clement (Fr).

Iridium (element) 1804. *Disc.* by Charles Tennant (1768–1838).

Iron *c.* 2250 B.C. Man-made iron in use as ornaments ceremonial weapons. *c.* 1400 B.C. The Chalybes, a subject tribe of the Hittites, *inv.* economic production of iron by the cementation process of steeling wrought-iron bars. Use of forge bellows implied. 4th cent. B.C. Cast iron in use in China. A.D. 200. Date of oldest known complete iron casting—a funerary cooking stove. 1249. A "moulin de fer" (iron mill) recorded in use in France. (The tilt-hammer probably *dev.* from the water-driven fulling-mill) (*q.v.*). 1320. Water-power applied to iron manufacture at Dobrilugk. 1709. Abraham Darby (1677–1717) *inv.* process of smelting iron by mineral fuel (coal). 1745. C. Polhem (Swed) (1661–1751) *inv.* puddling process for wrought iron using grooved rollers. 1766. First definite progress in making wrought iron made by Brothers. T. and G. Cranage, of Colebrookdale. 1783. Puddling and rolling processes *inv.* by Welshman Peter Onions, of Colebrookdale and also by Henry Cort (1740–800), of Fareham, Hants. 1818. S. B. Rogers, of Nantyglow, Wales, *intro.* iron instead of sand for bottom of puddling furnaces. 1830. Joseph Hall (1789–1862), of Tipton, Staffs. *intro.* iron oxides as bottom lining for puddling furnaces. (*Inv.* never *pat.*) 1833. Hot blast with anthracite *pat.* in U.S. 1837. Same process in use in Wales.

Iron Lung *See* Respirator, artificial.

Iron, Structural 1851. James Bogardus began cast-iron frame building in New York. 1851. The Crystal Palace, built to house the Great Exhibition erected after prefabrication of 12,000 iron girders, in Hyde Park, London, after designs of Joseph Paxman. 1889. Eiffel Tower *des.* and erected by Gustave Eiffel for Paris Exhibition of 1889. (In England churches built by John Cragg and Thomas Redman.)

Irrationals (mathematics) 6th cent. B.C. *Disc.* by Pythagorus (A.G.).

Isomorphism, Law of 1818. *Disc.* by Mitscherlich (1794–1863).

Isoperimeters (mathematics) Problem of solved by Jacques Bernouilli (1654–1705). (*See also* Calculus of variations.)

Isothermal Lines First delineated on map by Baron Alexander von Humboldt (Ger) (1769–1859). 1848–64. Theory of isothermal lines enunciated by Dove.

Isotopy (chemistry) Name coined by Frederick Soddy. 1913. First demonstrated that atoms of an element were not exactly identical made by Joseph John Thomson (1856–1940). 1923. Isotopy first used for investigating biological problems by Hevesy.

Ivory 1000 B.C. Ivory combs and trinket-boxes were made in Syria.

J

"J" (letter) 1550. *Intro.* into alphabet by Parisian printer Giles Beys.

Jack (Tool) *c.* 1250. Screw-jack sketched by Villard de Honnecourt (Fr) (*c.* 1245) and mentioned by monk Gervais (Fr).

"Jenny," Spinning 1745. Andre l'Aine, of St. Jean-en-Royans, Dauphine *inv.* spinning jenny for flax, cotton, wool and hemp; using three bobbins simultaneously. 1755. François Nicholas Brizout de Bainville (*d.* 1772), a Rouen merchant *inv.* machine which did work of 150 persons, making thread twice as fine as Indian muslin. 1864. Spinning jenny *inv.* by James Hargreaves.

Jews' Harp 1591. First mentioned.

Joint-wiping Process First *inv.* John and George Alderson. 1820. Process *imp.* by Thomas Dobbs, of Birmingham. 1837.

Further *imp.* by John Wainer, Jnr.

JOULE, James Prescott (1818–89), of Manchester. 1843. *Pub.* results of experiments in mechanical equivalent of heat—the kinetic theory. (*See also* Calorimeter.) Unit of heat named after him.

Juno (planet) 1804. *Disc.* by Karl Ludwig Harding.

Jupiter, Satellites of 1610. *Disc.* by Galileo Galilei (1564–1642).

Jute 1830. *Intro.* into England for sack-making.

K

Kaleidoscope 1817. *Inv.* Sir David Brewster (1781–1868).

Kamptulicon (floor covering) 1844. *Pat.* by Elijah Galloway (U.S.).

Kaolin (china clay) 1760. First *disc.* in England, in Cornwall, by William Cooksworthy.

KAPLAN, H. C. V. 1913. *Pat.* water turbine with adjustable rotor blades.

KAY, John (1704–64) 1733. *Inv.* the "flying shuttle."

KEPLER, Johannes (1571–1630) 1609. *Prop.* the laws of planetary motion in his "Astronomia nova."

Kerr Cell 1845. Michael Faraday (1791–1867) *disc.* that polarized light was diverted by

a magnet. 1877. John Kerr showed that polarized light was rotated by reflection from the end of a magnet. 1905. Cotton and Monton *assoc.* Kerr phenomenon with magnetic field (the "C.M." effect).

Keys *See* Locks.

Key, Telegraph Transmitting *c.* 1836. *Inv.* Prof. Samuel Finley Breeze Morse (U.S.) (1791–1872).

Kidney, Artificial *Inv.* Dr. W. J. Kolff.

Kinematoscope 1860. *Inv.* Colman Sellers (U.S.).

Kinetic Theory 1841. Waterson made first important attempt to formulate the theory.

Kinescope *See* Cathode ray tube.

Kinetoscope 1888. *Inv.* Thomas Alva Edison (U.S.) (1847–1941). *See also* Cinematograph.

Kite 1000 B.C. Originated in China. 1405. Hot-air kite *illus.* in German MSS. 1450. Dragontype kite *illus.* in Viennese MSS. 1589. Diamond-shaped kite *desc.* by Della Porta (1535–1615). 1752. Benjamin Franklin (U.S.) (1706–90) used kite to demonstrate electrical nature of lightning. 1804. Sir George Cayley used kite to form wing of his model glider—the first true aeroplane recorded. 1825. George Pocock recorded as making man-lifting kite. 1827. Pocock used kites for drawing his kite-carriage, or "char volant." 1859. Kite used by E. J. Cordner for ship-to-shore rescue work. 1893. Box kite *inv.* Lawrence Hargrave (Australia). 1894. Capt. B. F. S. Baden-Powell used a train of box-kites to lift a man. 1901. Col. S. F. Cody *pat.* man-lifting kite system. 1905. German kite carrying meteorological instruments reached height of 4 miles. 1907. Tetrahedral kite *inv.* Dr. Alexander Graham Bell. This kite was tested in form of an aeroplane in 1909. (Athanasius Kircher (1601–80), in his book *Ars Magna Lucis et Umbrae* (1646) alleges contemporary man-lifting kites were used in Rome.) *See also* Aircraft.

KLAPROTH, Martin Heinrich (Ger) (1743–1817) *Disc.* metals titanium, zirconium and uranium (*q.v.*).

Knitting Machine 1589. *Inv.* by Lee. 1758. *Imp.* by Jebediah Strutt.

Knives 1563. First made in England. 1600. Clasp-knife *intro.* from Flanders.

Knife-cleaning Machine 1844. *Inv.* by George Kent.

Knot (nautical measure) 1607. First mention in book on an East Indian voyage.

Krypton (elemental gas) 1898. *Disc.* by Prof. William Ramsay (1852–1916).

KÜNKEL, Johann (Baron Lowenstjern) (Ger) (1630–1703) *Disc.* phosphorus.

Kymograph *Inv.* Karl Ludvig (1816–95) for mechanically recording on a revolving drum a permanent record of continuous movement (self-recording instruments).

L

LACAILLE, Nicholas-Louis (Fr) (1713–62) Astronomer. Catalogued 2,000 fixed stars and determined their proper motions.

Lace 1320. First made in France and Flanders. 1626. First made in England at Great Marlow, Bucks. 1829. First made in Ireland at Limerick.

Lace-making Machine 1758. *Inv.* by Strutt. 1809. Bobbin-net machine *inv.*

Lacquer *c.* 1650. *Inv.* in imitation of much earlier Chinese and Japanese lacquers by D'Agly, of Liege. 1744. Process of lacquering *imp.* by S. E. Martin.

Lacteal Vessels (anatomy) 1627. *Disc.* Gasparo Aselli (It) (1581–1626).

Lactic Acid 1780. *Isol.* from sour milk by C. W. Scheele (1742–86). 1832. Found in meat extract by Baron von Liebig (1803–73). 1850. Synthesized. 1857. Formation in milk ascribed by Louis Pasteur (1828–95) to micro-organisms. 1872. Two types of lactic acid *disc.* by J. Wislicenus one of which was optically active.

Lactometer Pre-1835. *Inv.* by Dicas, of Liverpool.

Ladder, Portable 1655. Forms subject of Marquise of Worcester's *Century of Inventions, No. 50.* *c.* 1830. Portable ladders had been *pat.* by William Hilton,

Gregory, and Green, of London.

LAGRANGE, Joseph-Louis, Comte de (Fr) (1736–1813) Originator of the calculus of variations. *Prop.* general principles of dynamics and dynamic analysis. *Disc.* "Lagrangian Point" in orbit of planetary body.

LALANDE, Joseph-Jérôme de (Fr) (1732–1807) 1760. Determined parallax and distance of the moon at 57° 15', corresponding to 384,000 km.

LAMBERT, Johann Heinrich (Ger) (1728–77) 1761. *Prop.* first theory of the Milky Way.

Lamb-shift (nuclear physics) 1947. *Disc.* by Willis E. Lamb (U.S.) in collaboration with R. C. Rutherford. (Subject later elucidated by Kramers (Hol), H. A. Bethe (Alsace), and Julian Schwinger (U.S.)

LAMONT, Johann von (Ger) (1805–79), of Munich. 1851. *Disc.* relationship between sunspot activity and the earth's magnetic field.

Lamps (all types) 890. Lamps with sides of scraped horn used by King Alfred. 1415. London streets first lit by lanterns. 1681. London streets first lit by suspended oil lamps. 1683. Otto von Guericke (1602–86) obtained light from discharge of static electric machine. 1700.

Isaac Newton (1642–1727) and Francis Hauksbee (1687–1763) produced light from exhausted glass spheres with frictional electricity. 1710. Air in glass tube made to glow by electricity by Hauksbee. 1744. Use of electric discharge lights proposed for German mines. 1744. Reflecting oil-lamp *inv.* by Bourgeois de Chateaublanc and used in streets of Paris. 1780. Aimée Argand (Swit) (1755–1803) *inv.* his wick oil-lamp. 1800. Guillaume Carcel (Fr) (1750–1812) *inv.* pressurized oil-lamp. 1802. Electric arc light *disc.* Sir Humphry Davy (1778–1829). 1806. Safety lamp conceived by Dr. Clauny. 1814. London streets first lit by gas. 1815. Miners' safety lamp *inv.* Sir Humphry Davy, 1820. (some authorities, 1809). W. De la Rue *disc.* that electric current passing through platinum wire coil in partial vacuum produced light. Pre-1825. Moderator oil-lamp *inv.* Le Vavasseur, Hadrot, and Neuburger. 1841. F. De Moleyns took out first British *pat.* for incandescent electric lamp. (Powdered charcoal between platinum coils in a vacuum vessel.) 1845. J. W. Starr, of Cincinnati, U.S. *inv.* first incandescent electric lamp; stating that "when a carbon filament is used, it should be enclosed in a vacuum." (This *pat.* was followed up by Swan many years later.) 1849. Leon Foucault (Fr) (1819–68) *inv.* clockwork-magnetic control for electric arc-lamps. 1850. Submarine lamp *inv.* by Siebe and Gorman. 1856. German glass-blower-artist Geissler originated electric discharge tube run on high-voltage alternating current in either air or gas-filled tubes. 1857. Becquerel experimented with electric gas-discharge lamps coated with luminescent materials. 1858. Carbon arc-lamp *inv.* 1860. Joseph Wilson Swan (1828–1914) made electric vacuum lamps with "U"-shaped filaments of carbonized paper. Mercury-vapour lamp *inv.* by Prof. Way and tried at Osborne House, Isle of Wight. (Pressurized mercury column in vacuo.) 1860 (Sept.). Hungerford suspension bridge, London, illuminated by Way's lamps. 1862. Timothy Morris, Edward Moncton and Robert Weare jointly took out first British *pat.* for electric discharge-tube lamp. 1866. Adolphe Miroude *inv.* battery-operated, nitrogen-filled tube for buoy lighting. 1873. First permanent installation of electric light at Paris workshops of Z. T. Gramme (1826–1901). 1877. C. W. Brush (U.S.) makes his first arc-lamp. Telegraph engineer Paul Jablochkoff (Rus) (1847–94) *inv.* "electric candle." 1879. T. A. Edison (U.S.) (1847–1931) made a platinum filament vacuum lamp and others with Bristol-board and cotton (charred) filaments. 1880. Edison *pat.* electric lamp with bamboo filament. 1888. Presence of certain impurities in luminescent substances used in gas-discharge lamps known to be necessary for their proper functioning. 1891. Philips commenced mass-production of vacuum lamps in Holland. 1897. Von Bolton and C. A. Welsbach succeeded in making vacuum lamp with malleable tantalum filament. 1901. Peter Cooper-Hewitt (U.S.) *inv.* first low-pressure, mercury-vapour gas-discharge lamp. 1903. Hanemann and Just *dev.* tungsten filament vacuum lamp. 1904. Inside frosting of vacuum lamps *inv.* 1904. Klatt and Lenard investigated impurities in gas-discharge lamps. 1905. Osmium filament first used in vacuum lamps. 1906. High-pressure, mercury-vapour quartz lamp *inv.* Küch (Ger). 1907. "Osram" (tungsten osmium) filament lamp *intro.* 1908. Tantalum filament lamp *intro.* 1909. Tungsten filament vacuum lamp *intro.* 1913. Coiled tungsten filament in argon gas-filled bulb *inv.* Dr. Irving Langmuir (U.S.). 1923. Use of rhodamine proposed to improve light from mercury lamps. 1926. Discharge-lamps with fluorescent materials within bulb *inv.* Edmund Germer, Friederich Mayer and Hans Spanner (all Ger). 1930. Foil-type electric photo-flash lamp *inv.* by Johannes Ostermayer (Ger). 1934. Argon gas-filled, "coiled-coil" tungsten filament lamps *intro.* 1935. "Photo-flash" type photographers' flash-lamp *dev.* by engineers of N. V. Philips, Holland. 1935. Bright yellow light under illumination from neon lamp *prod.* by H. G Jenkins by use of zinc ortho-silicate and calcium tungstate. (*See also* Gas lighting and Lighting systems, Motor-car.)

Lamp, Safety 1814. W. R. Clanny (1776–1850) *inv.* his "blast lamp." 1816. Sir Humphry Davy *inv.* his third safety lamp. 1840. Davy's lamp *imp.* by Meuseler (Bel), and L. Marsant (Fr).

LAMPADIUS, Wilhelm Augustus (Ger) (1772–1842) 1811–16. Successfully experimented with gas-lighting (*q.v.*) at Freiburg.

LANGLEY, Samuel Pierpoint (1834–1906) 1880. *inv.* the bolometer.

Language, Universal 1661. Mentioned by the Marquis of Worcester in his *Century of Inventions, No. 32.* 1661. *Inv.* by John J. Becher (1635–82).

Lanitol (artificial wool fibre) 1924. *Prod.* from casein by Antonio Ferretti (It).

Lanthanum (element) 1839. *Disc.* Carl Gustav Mosander (Swed).

Lapis Lazuli, Artificial 2000 B.C. Made in Egypt from sand, chalk, soda and malachite.

LAPLACE, Pierre Simon Marquis de (Fr) (1749–1827) Instituted calorimetric experiments. Calculated irregularities in motions of planets Jupiter and Saturn.

Laryngoscope 1855. *Inv.* by Manuel Garcia (1805–1906), a Parisian singing-master. 1857. *Imp.* by Dr. Czermak, of Pesth.

Laser (Acronym = Light Amplification by Stimulated Emission and Radiation) (1878. Graham Bell and Sumner Tainter *pat.* 40 ways of transmitting telephony along a light-beam.) 1958. Maser (*q.v.*) techniques suggested to be extended to visible light by A. L. Schawlow and C. H.

Townes. 1960. Schawlow's and Townes' theories demonstrated by Theodore H. Maiman, using a pink, synthetic ruby crystal of sapphire doped with .05 per cent trivalent chromium. 1961. Multilevel Laser with uranium or samarium in calcium fluoride announced. 1962. Semi-conductivity Laser using gallium arsenide *disc.* by R. N. Hall, of Schenectady, N.Y. The first continuously operating Laser operated by radio-frequency discharge in helium/neon gas *inv.* by A. Javan. (*See* Maser.)

Latex 1933. First foam latex sponge *prod.* by D. F. Twiss.

Lathe 600 B.C. *Inv.* ascribed to Talus, grandson of Daedalus; and by Pliny to Theodore of Samos. 400 B.C. Mentioned by Plato and Virgil. Pole lathe used since 13th cent. Treadle lathe with flywheel mentioned by Leonardo da Vinci (1452–1519). 1568. Screw-cutting lathe *inv.* by Dauphinois mathematician Jacques Besson, of Besangon (Fr) (*c.* 1550). 1579. Oval-turning lathe first mentioned by Besson. 1648. Maignan *pub.* in Rome engravings of two lathes for turning metallic optical mirrors to form hyperbolic, spherical, or plane surfaces. 1710. Christopher Polhem built lathe of capacity to machine work of industrial sizes. Engineer J. Fortin (Fr) (1750–1831), also responsible for early *dev.* of lathe. 1772. First note of the slide-rest. 1794. Joseph Bramah (1748–1814) *inv.* (?) slide-rest. 1804. Medallion lathe suggested by James Watt (1736–1819). 1818. Copy-

ing lathe *inv.* Thomas Blanchard (U.S.). 1820. Henry Maudslay (1711–1832) *inv.* (?) slide-rest lathe. 1820–30. Medallion lathe independently *inv.* by Benjamin Cheverton (U.S.). 1855. Turret lathe (screw milling machine) *inv.* by F. W. Howe, Richard Lawrence, and Henry D. Stone, of Robbins and Lawrence, small-arm makers, of Windsor, Vermont, U.S. 1857. Turret block first made automatic. 1857. Taylor and White make machine lathe universal for mass-production. 1862. Turret lathe *intro.* into Britain. 1891. Flat turret lathe *inv.* by James Hartness.

LAURENT, Augustin (1807–53) 1836. *Form.* the so-called nucleus theory by which substitution of atoms could yield a series of chemical compounds all connected by an unaltered "nucleus."

LAVAL, Carl Gustaf Patrik de (Swed) (1845–1913) 1887. *Pat.* steam turbine. 1890. *Inv.* helical gears.

LAVOISIER, Antoine-Laurent (Fr) (1734–94) 1774. First to explain chemical composition of air, and why things burn. Named the gas "oxygen," having *disc.* it.

LAWES, Sir John Bennet (1814–1900) 1843. Established fact that nitrogen, potassium and phosphorus were the elements most needed as fertilizers (*q.v.*).

Lead (metal) *Disc.* pre-1000 B.C. A.D. 714. Mines at Wirksworth, Derbyshire in production. 1790. Ironmaster John Wilkinson *pat.* process for lead pipe-casting. 1797. First lead pipes cast by extrusion. 1822. Continuous pipe-casting process *pat.* John Hague. 1827. Lead ore-washing process *imp.* by Harsleden. 1830. *Imp.* furnaces for smelting lead *inv.* Joseph Wass, of Ashover, Derbyshire. 1919. Process for softening lead *inv.* by Harris. 1920. Parkes's de-silvering process adapted to be continuous by G. K. Williams (Australia).

LE BELL, Joseph Achill (1847–1930) 1874. With van't Hoff *prop.* a consistent theory of molecular structure based on the spatial arrangement of the four valencies of the element carbon.

LE CHATELIER, Henry (1850–1936) *Inv.* a reverberatory gas furnace with air intake pre-heated by flue gases, to produce sufficient heat to make steel (*q.v.*).

LE FEBVRE, Nicaise (Fr) (1620–74) Pioneer of the Paracelsan chemistry school who pioneered the modern textbook of chemistry by writing a chemico-pharmaceutical handbook in 1670.

LE MONNIER, Louis-Guillaume (Fr) (1717–99) *c.* 1750. Made important researches into atmospheric electricity.

LEBLANC, Nicholas (Fr) (1742–1806) 1787. *Inv.* process for making sulphate of sodium by decomposing sodium chloride with sulphuric acid. 1825. Leblanc process first used in G.B. by Charles Tennant. (*See also* Bleaching power.)

LEBON, Phillippe (Fr) (1769–1804) 1799. *Inv.* early engine to run on coal-gas. *See* Engine, gas.

LECLANCHÉ, Georges (Fr) 1839–82) 1867. *Inv.* the electric battery now bearing his name.

LEEGHWATER, Jan Adrianszoon (Hol) (1575–1650) Civil engineer who first submitted plan to drain the Haarlemermere, using 100 windmills for the purpose.

LEEUWENHOEK, Antonie van (Hol) (1632–1723), of Delft. Naturalist and microscopist. Founded science of histology. 1683. First to see bacteria with a microscope of his own construction. *Disc.* compound eyes of insects.

LEIBNITZ, Gottfried Wilhelm (Ger) (1646–1716) 1684. Created the differential and integral calculus, which was added to by Jacques Bernoulli (1654–1705) and his brother Jean Bernoulli (1667–1748), and by Jean's son Daniel (1700–82) (*q.v.*).

LEIDENFROST, Johann Gottleib (Ger) (1715–94) *Disc.* the spheroidal state which occurs when water is dropped on to a red-hot metal plate.

LEMERY, Nikolaus (Fr) (1645–1715) 1677. Pioneered the modern chemical text-book. (*See also* Nicaise Le Febvre.)

LENOIR, J. J. Etienne (Fr) (1822–1900) 1860. *Pat.* gas-engine with electric coil ignition. Working on the two-stroke principle, this engine was the first commercially successful gas-engine (*q.v.*).

LEONARDO da Vinci *See* Vinci, Leonardo da.

Lens Originally of quartz or beryl (aquamarine). There is a lens from Ancient Assyria in the British Museum. A.D. 2–65.

Seneca's writings show that he knew of the magnifying power of glass in convex form.

Lens, Achromatic 1695. Combination of different refracting media to produce achromatism suggested by Sir Isaac Newton's pupil James Gregory (1638–75). 1733. First achromatic lens perfected of flint and crown glass by Chester Moor Hall (1703–71). 1755. Commercial production of achromatic lenses commenced by John Dolland (1706–61). 1870. Karl Zeiss and friend Abbé founded Jena optical glass works—first world centre of optics.

Lens, Contact 1887. First made by A. E. Fick. 1938. Plastic contact lenses of methylmethacrilate used. 1950. Smaller contact lenses used, covering cornea only. 1958. Bifocal contact lenses *intro*.

Leprosy Bacillus of leprosy *disc*. Armauer Hansen, of Bergen (1841–1912).

LESLIE, Sir John (1766–1832) 1800. *Inv.* hygrometer. Also *inv.* differential thermometer, a photometer, and a method of freezing by rapid evaporation.

LEUCIPPUS (A.R.) (6th cent. B.C.) Inaugurated the philosophy bearing his name.

Leucin (inorganic chemistry) *Disc.* in urine by Friederich von Frerichs (Ger) (1819–85).

Leucocythemia (medicine) 1845. First *disc*. by John Hughes Bennet (1812–75), of Edinburgh.

LEUPOLD, Jacob (Ger) (1674–1727) Wrote several books on hydraulics, fire-engines, and machines in general; as well as a description of a proposed high-pressure pumping engine with two single-acting cylinders. This type of machine not used until made by Cugnot in 1769 (*q.v.*).

Lever, Laws Relating to the 287–212 B.C. *Disc.* by Archimedes.

Level, Surveyors' 1737. *Inv.* Abbe Soumille (Fr). 1802. *Imp.* by Huette (Fr).

LEVASSOR, Émile (Fr) (1814–97) Pioneered the commercial application of Gottleib Daimler's internal combustion engine of 1883–85 to a road vehicle. (*See also* Engine, internal combustion.)

LEVERRIER, Urbain, Jean-Joseph (Fr) (1811–77) Explained (with John Couch Adams, of Cambridge) the irregularities found in the motion of the planet Uranus as the result of a hypothetical planet; the presence of which was established on Sept. 23, 1846 by John Gottfried Galle, of Berlin. New planet later named Neptune.

Leyden Jar (condenser) 1745. Independently *disc*. Ewald Georg von Kleist, at Kamin, Pomerania, and in 1746 by Petrus van Musschenbroeck, a German pastor and scientist living in Leyden. *See also* Condenser, Electrical.

LIBAVIUS, Andreas (1540–1616) 1597. Gave an exact description of chemists' apparatus and all operations such as calcination, sublimation. His classification of bitumens, natural waxes and resins still stands, in principle, today.

LIBBEY, Edward Drummond (U.S.) (1854–1925) 1899–1904. With M. J. Owens, of Toledo, *dev.* the first glass-bottle casting-machine.

LIEBIG, Justus Freiherr, Baron von (Ger) (1803–73) 1823. *Disc.* that silver fulminate and silver cyanate have same percentage composition, but entirely different chemical properties. *Dev.* gravimetric and volumetric techniques of chemical analysis. 1834. Proved the presence of the radical "ethyl" in ether and ordinary alcohol. 1850. *Inv.* process for making meat extract by evaporation.

Lifeboat 1785. Lukin granted first English lifeboat *pat.* 1789. Henry Greathead, of South Shields *inv.* lifeboat. 1853. H. F. Richardson *inv.* tubular lifeboat. G. F. Barrett *inv.* expanding tubular life-raft. 1890. Greathead's lifeboat awarded £12,000 by Parliament. Self-righting lifeboat *inv.* by William Wouldhave, of South Shields.

Life-saving Apparel 18th cent. (late). Le Conte (Fr) *inv.* air-filled life-saving vest.

Lifts (elevators) 236 B.C. Archimedes *des.* and *const.* a lift. 1850. Platform for hoisting barrels made and installed in New York by Henry Waterman. 1853. Elisha Graves Otis (U.S. *des.* and *prod.* his first lift with automatic safety devices. 1857. Otis erected lift with totally enclosed car at Haughwort's Store, New York. 1859. Fifth Avenue Hotel, New York, first hotel to be equipped with lift. 1866. Steam-driven lift installed at St. James's Hotel, New York. 1867.

Felix-Léon Edoux (Fr) *inv.* and installed hydraulic lift at Paris Exposition. 1868. Equitable Life Insurance Society the first office building to be equipped with a lift. 1889. Otis *des.* and *pat.* first electric lift.

Light 1100. Refraction of light *disc.* Alhazen (Arab). 1621. Laws of refraction of light formulated by Willebrord Snell. 1664. Sir Isaac Newton *disc.* composition of white light by means of a prism and named the "spectrum." Diffraction of light *disc.* Grimaldi (1618–63). 1692. Polarized light *disc.* by Christian Huygens. 1727. Abberation of light of fixed stars *disc.* by James Bradley (1692–1762). Undulatory theory of light proved by Augustin-Jean Fresnel (Fr) engineer (*c.* 1820). (His hypothesis advanced by Huygens and further elaborated by Euler; being finally completed by Thomas Young and substituted for Newton's corpuscular theory.) 1808. Polarized light *disc.* by Étienne-Louis Malus (Fr) (1775–1812).

Light, Wave-lengths of Early 19th cent. Thomas Young and Joseph Fraunhofer invented simple devices for measurement of. *c.* 1868. Ångström measured wave-length of light. 1883. Rowland measured wave-length of light more accurately.

Light, Law of Refraction of 1621. *Disc.* by Snell (1591–1626) (Snell's Law).

Light, Speed of 1676. Estimated by Röemer at 192,000 miles per second. 1849. Determined by Armand-Hippolyte-Louis Fizeau (Fr) at 190,000 miles per second, using a rotating toothed wheel. 1850. Jean-Baptiste-Léon Foucault (Fr) measures speed of Light with revolving mirror. 1880. Speed of light checked by Simon Newcombe (U.S.). 1882. Speed of light checked by Albert Michelson (U.S.) (both of Cleveland, Ohio).

Light, Infra-red Rays of 1800. *Disc.* by William Herschel in the spectrum of the sun.

Light, Lime- Dr. Robert Hare (U.S.) (early 19th cent.) suggested hydrogen burning in oxygen to produce a greater heat than any previously known. 1826. *Inv.* by Capt. Drummond. 1839. *inv.* Sir Goldsworthy Gurney, of Bude, Cornwall. (*Pat.* 1841 and known as the "Bude Light.")

Lighthouse 1840. First cast-iron lighthouse erected at Morant Point, Jamaica. 1856. Henry Hale Holmes *pat.* electric generator used to light South Foreland lighthouse (first electrically lighted lighthouse).

Lighting Systems and Lamps (carriage and automobile) 1785 and 1793. Joseph Lucas *pat.* oil lamps with convex mirrors and lenses. 1896. Louis Bleriot (Fr) *inv.* self-contained acetylene gas lamp. 1905. Lucas makes his first acetylene gas lamp. 1905. Besnard, H. Salsbury, and T. Whitaker *pat.* anti-dazzle electric car lamp. 1924. W. T. G. Fenton and J. Riddington *pat.* dipping motor-car headlamp. 1924. Double-filament electric lamp bulb for motor-car use *inv.* by A. Graves. *c.* 1910. C. A. Vandervell and W. H. Proctor perfect electric motor-car lighting system with automatically controlled dynamo. 1937. E. A. Howard (U.S.) and K. D. Scott (U.S.) *inv.* "sealed-beam" motor-car headlamp.

Lightning Conductor 1752. *Inv.* by Benjamin Franklin (U.S.) (1706–90), a Philadelphia printer, and first demonstrated at Marly, France by D'Alibard. (Franklin had already established the relationship between electricity and lightning in 1749.)

Limbs, Artificial 1570. Breton nobleman and Hugenot soldier François de la Noué used an iron arm. 1600s. Jean Truchet and Du Quet, of Paris, made movable artificial hands for a Swedish soldier.

LINDE, Karl Paul Gottfried von (Ger) (1842–1934) 1895. One of the pioneers of low-temperature gas liquefaction. (*See also* Cryogenics.)

LINNAEUS, Karl (Swed) (1707–78) 1735. Assigned class, order, genus, and species to every known animal and plant in his "Systema Naturæ."

Linotype Machine 1873. *Inv.* by 18-year-old watchmaker, Ottmar Mergenthaler, of Washington, D.C., U.S. 1886. First commercial linotype machine installed in New York. 1892. *Inv.* pioneered in Europe by Sir Joseph Lawrence. 1892. A new model linotype machine appeared in Britain.

LIORSO, A. G. (U.S.) *See* Seaborg, Glen T.

Lipoid (chemistry) 1862. Term first used by E. Wagner (Ger).

LIPPERSHAY, Johannes (Hol) (*d.* 1619) Lens-grinder of Middelburg, credited with the *inv.* of spectacles and the telescope (*q.v.*).

Liquation (metal recovery process) Used by Ancient Romans in production of gold and silver, which were melted in copper and lead; the metal in the copper being transferred to the lead, whence it was recovered by cupellation. (*See also* Amalgams.)

LISTER, Dr. Joseph, Baron (1827–1912) 1865. First used an antiseptic at an operation.

LISTON, Dr. Robert (1794–1847) *Inv.* Surgical long splint and bone forceps.

Lithium (Lithia) (element) 1817. Lithium *disc.* by Arfvedson, a pupil of Berzelius. Lithium finally *isol.* electrolytically by Sir Humphry Davy.

Lithography 1796. Process *inv.* by musician Aloysius Senefelder (1771–1834), of Prague. 1799. Process *pat.* 1801. Process *intro.* England by André d'Offenbach. 1810. Hullmandel erects first lithography press in London.

Lithotrite (surgery) 1824. *Inv.* by Jean Civiale (Fr).

Liver, Function of the *Disc.* by Claude Bernard (1813–78), who also *disc.* functions of the pancreas, salivary gland and spinal cord.

LJUNGSTRÖM, B. (Swed) (1872–1948) Made important *invs.* relating to the application of the steam turbine to the locomotive. (*See* Turbo-locomotive and Turbine.)

Locks (and Keys) *c.* 2000 B.C. Egyptian mummy-case locks with weighted tumblers in use. 730 B.C. Earliest mention of locks by Homer in *Odyssey*, xxi, with "brass key." A.D. 800. Lock of this date found with key at Thebes. 1320. Lock on door of church at Gedney, Lincolnshire. 1540. Pad-locks "*inv.*" in Nuremburg. 1682. "Letter-lock" or combination padlock *desc.* in book printed at Amsterdam. 1739. Cornthwaite *inv.* lock. 1774. Barron *pat.* lock with two or more tumblers. 1784. Joseph Bramah *inv.* lock eventually picked by Alfred Hobbs in 51 hours, spread over 16 days. 1790. Moses Bird *inv.* the lever lock. 1792. Watchmaker Santos le Gendre *imp.* safety lock. 1816. Kemp, of Cork *pat.* "Union" lock, a combination of Barron's and Bramah's. 1818. Jeremiah Chubb *inv.* his first lever lock. 1831. William Rutherford, of Jedburgh *inv.* lock with stopplate. 1834. Chubb lock simplified. 1846. Cover-plate *inv.* De la Fons. 1850s. Andrews, of Perth, Amboy, U.S. and Newell, of New York, *inv.* permuting locks. 1852. Hobbs *inv.* "Protector" lock. 1865. "Yale" pin-tumbler lock *inv.* Linus Yale, Jnr. (U.S.). (This was first time mass-production methods were used to produce unidentical articles.)

LOCKE, John (1632–1704) One of the pioneers in the investigation of heat and its effects.

Lock Gates 1495. *Inv.* in the West by Leonardo da Vinci, who, in 1497, used mitred lock gates on the Milan Canal. 1545. In use in France, which then had 2,215 miles of canals. 1564–67. First used in England on the Exeter Canal.

LOCKYER, Sir Joseph Norman (1836–1920) *Inv.* prism camera for astronomical spectroscopy.

Locomotive, Articulated John Cockerill, of Seraing, Belgium, *des.* articulated locomotive "Seraing" for the locomotive trials at Semmering. 1863. Robert Francis Fairlie (Scot) *pat.* articulated locomotive "Progress" for the Neath and Brecon Railway. 1869. William Mason (U.S.) modified Fairlie's design. 1887. Pechot-Bourdon (Fr) built an articulated locomotive similar to Fairlie's. 1885. Johnstone (U.S.) builds articulated locomotives in U.S. at his Rhode Island Works. 1887. Anatole Mallet (Bel) built his first articulated locomotive at Atliers Metallurgiques, Turbize.

Locomotive, Diesel *See* Engine, diesel.

Locomotive, Diesel-Steam 1920. *Inv.* by W. J. Still, of Still Engine Co., Chiswick, London, in conjunction with Lieut. Col. E. Kitson-Clarke, of Messrs. Kitson, of Leeds. Locomotive built in 1927.

Locomotive, Electric 1842. Thomas Davenport, of Brandon, Vermont, U.S., propelled a coach 16 ft. long by 6 ft. wide, weighing 5 tons on the Edinburgh and Glasgow Railway at 4 m.p.h. (*See* also Railway, electric.)

Locomotive, Ice 1862. Ice locomotive "Rurik" *des.* by Nathaniel Grew for use on the River Neva between St. Peters-burg and Kronstadt, when frozen over.

Locomotive, Petrol-driven *c.* 1899. Tried on Würtemburg State Railways.

Locomotive, Steam Railway 1803. Richard Trevithick, of Illogan, Cornwall *inv.* first high-pressure locomotive, which was tried successfully on the 9 mile Pen-y-Darren Tramroad in Wales. 1808. Trevithick's 10-ton locomotive "Catch-me-who-can" achieved a speed of 12–15 m.p.h. on a circular track in Euston Square, London. 1811. John Blenkinsop, of Leeds *des.* rack-and-rack-rail locomotive. 1812. Chapman *inv.* locomotive which hauled itself along a cable, also *pat.* four-wheeled bogie (for use on coal wagons). 1813. Christopher Blackett and William Hedley, of Wylam Colliery, Newcastle upon Tyne pirated and *pat.* Trevithick's *inv.* 1813. William Brunton *pat.* a locomotive propelled by steam-operated pusher-legs. 1814. George Stephenson *des.* and builds his first locomotive—"Blucher." 1815. Stephenson, in conjunction with Dodd and Losh *pat.* outside connecting and coupling-rods. 1824. Stephenson's engine "Locomotion," with coupling-rods opened the Stockton and Darlington Railway. 1827. Timothy Hackworth rebuilt the S. & D.R. engine "Royal George" with an exhaust blast-pipe. It could haul 130 tons on the level at 5 m.p.h. Marc Seguin (Fr) *pat.* multi-tube locomotive boiler. 1828. Stephenson *des.* "Lancashire Witch" for the

Bolton and Leigh Railway, fitting each piston-rod with a crosshead and slide-bars. 1828. Foster, Rastrick and Company build first locomotive for use in the New World—the "America." "Stourbridge Lion," however, was the first locomotive to be tried there. 1829. Locomotives "Novelty," "Sanspareil," and Stephenson's "Rocket" compete in the Rainhill locomotive trials, in which "Rocket" hauled 30 passengers at 30 m.p.h. 1829. Stephenson *des.* his "Planet" class locomotive. 1830. Hackworth *des.* the first locomotive with inside frames and inside cylinders— the "Globe." ("Planet" and "Globe" were the locomotives from which modern engines have been derived, containing horizontal cylinders enclosed within the smoke-box, cranked driving-axles, multi-tubular boilers, and axle-boxes running in horn-plates carrying the running-wheel axles, together with adequate springing.) 1830. Peter Cooper (U.S.) builds locomotive "Tom Thumb." E. L. Miller *des.* and *const.* 2–2–0 + 0–2–2 locomotive "Best Friend of Charleston." 1831. Horatio Allen *des.* "South Carolina." 1831. David Matthew (U.S.) builds "De Witt Clinton." 1832. Ex-watchmaker Matthias Baldwin (U.S.) builds his first locomotive—"Old Ironsides." 1832. First U.S. locomotive with four-wheeled leading bogie. Stephenson began building six-coupled, inside-cylinder locomotives. 1833. First 0–6–0 and first bogie 4–2–0 shipped to U.S. 1834. John G. Bodmer *inv.*

self-balanced, double-piston locomotive. 1835. Solid wrought-iron wheels *inv.* John Day. 1837. William Fernihough applied counterbalance weights to locomotive drivingwheels. 1839. Isaac Dodds *inv.* non-strain locomotive boilermounting method, and wedgemotion. valve-gear. Hackworth's locomotive "Arrow" reached speed of 42.8 m.p.h. 1840. Stephenson *inv.* steam dome. 1841. Samuel Hall *inv.* smoke-consuming firebox. 1842. Stephenson-Howe valvegear *pat.* Baldwin (U.S.) *inv.* flexible-wheelbase locomotive. 1846. Cast-iron wheels for locomotives *pat.* by Brothers Sharp (previously fitted by Hackworth to "Royal George." 1846. Stephenson *intro.* three-cylinder locomotive. 1847. Trevithick *des.* locomotive "Cornwall." 1848. Ross Winans (U.S.) *intro.* "Camel" type locomotive. 1848. T. R. Crampton's unique locomotive "Liverpool" appears. 1850. Great Western Railway locomotive "Lord of Isles" appeared. James Samuel *des.* continuous expansion compound locomotive. 1855. Daniel Gooch *intro.* 4–4–0 type with 7 ft. drivingwheels. 1855. Matthew Kirtley and Charles Markham *des.* modern-type firebox. 1858. Giffard steam injector first fitted to English locomotive. 1860. John Ramsbottom *inv.* water pick-up gear—first laid on Chester and Holyhead Railway. Coal instead of coke began to be generally used for locomotives. 1861. John Haswell *inv.* balanced locomotive "Duplex." 1870. Patrick Stirling *des*. Great Northern Railway single-driver locomotive "No. 1." 1873. James Stirling *inv*. steam reversing-gear. 1877. Anatole Mallet (Bel) *inv*. satisfactory method of compounding. 1878. Francis Webb *des*. experimental locomotive on Mallet's system—tried in 1882. Egide Walschaertz (Bel) *inv*. outside valve-gear. 1888. Flaman *des*. triple-boilered locomotive for Belgian State Railways. 1894. Flaman and Salomon (Fr) *des*. twin-boilered locomotive for French Est Railway. 1898. H. A. Ivatt *des*. 4-4-2, "Atlantic" locomotive for Great Northern Railway. 1906. Schmidt superheater first fitted to English locomotive. G. J. Churchward *intro*. topfeed to boiler. 1908. Tenwheeled "Decapod" locomotive *intro*. on Great Eastern Railway. 4-6-2 "Pacific" type locomotive "Great Bear" *intro*. on Great Western Railway. 1921. Sir Nigel Gresley *intro*. "Pacific" type on Great Northern Railway. 1923. C. B. Collett *intro*. "Castle" class on Great Western Railway. 1924. R. M. Deeley *intro*. famous "Compound" on Midland Railway. 1926. R. E. L. Maunsell *intro*. "Lord Nelson" on Southern Railway. 1927 C. B. Collet *intro*. "King" class on Great Western Railway. 1930. R. E. L. Maunsell *intro*. "Schools" class on Southern Railway. 1933. Sir William Stanier *intro*. "Princess" class on London, Midland and Scottish Railway. 1935. Stanier *intro*. "Jubilee" class. 1938. Stanier *intro*. "Corona-

tion" class. 1941. O. V. Bulleid *intro*. "Merchant Navy" class on Southern Railway. 1951. British Railways locomotive "Britannia" appears. 1954. British Railways locomotive "Duke of Gloucester" (4-6-2) appears. 1960. Last steam locomotive to be built for British Railways—"Evening Star," No. 92220, left Swindon works on March 18, 1960 and was scrapped in 1965. (*See also* Locomotive, steam-electric; Locomotive, diesel-steam; Locomotive, electric; Locomotive, turbine-driven, etc.)

Locomotive, Steam-Electric 1892. Prof. Heilmann (Ger) *des*. reciprocating steam-electric locomotive with a Brown steam-engine and Gramme dynamo and motors (*q.v.*). 1909. Reid-Ramsey steam-electric locomotive built at North British Locomotive Works. 1938. General Electric Co. (U.S.) *des*. and built 5,000 h.p. turbo-electric locomotive. 1947. Baldwin-Westinghouse steamelectric locomotive built.

Locomotive, Turbine-driven 1925. Krüpp-Zöelly turbine-driven locomotive built in Swit. B. Ljungström (Swed) also made important *invs*. relating to turbine-driven locomotive. 1935. Sir W. Stanier *des*. turbine-driven locomotive which ran on the L.M.S. Railway for 17 years.

Locomotive, Side-bevel-drive 1880. System *dev*. Ephraim Shay, of Lima, Ohio, U.S. 1894. Heisler locomotive *dev*. at Erie, Pennsylvania, U.S.

Log, Ship's Perpetual 1607. James Hookey's spinning log

first noted by Samuel Purchase. 1719. Log *inv.* by Pourcheff (Fr). 1772. Log *inv.* by William Foxon, of Deptford. 1802. Log *inv.* by Edward Massey. 1878. Walker *inv.* "Cherub" log.

Logarithms Steifel (1486–1567) *prop.* germ of idea, using letters for unknown numbers and signs + and −. 1614. John Napier, Baron of Murcheston (1550–1617) *dev.* and perfected logarithms and *intro.* word "Logarithm." 1617. Napier *pub.* first table of common logarithms. 1619. John Speidel *pub.* tables of logarithms of sines, tangents, and secants for every minute from 0° to 90° to five places of decimals. 1624. Henry Briggs (1561–1631) computed logarithms of numbers from 1 to 20,000 and 20,000 to 90,000 to 10 places of decimals. 1628. Dutch bookseller Adrian Vlacq computed logarithms from 20,000 to 90,000.

Logwood 1662. First *disc.* in Honduras by English.

LOMONOSOV, Mikhail Vasilyevich (Rus) (1711–65) Scientist who defended kinetic theory of pressure and heat experimentally established by Amontons.

Loom 4400 B.C. Horizontal ground loom of this date *disc.* at Badari, Egypt. 2000 B.C. Loom depicted on Theban tombs of this date. 1131. *Intro.* into England by the Flemings. 1600. Claude Dangon, of Lyons, *imp.* drawings of loom. 1604. Dutch "Engine-loom" *inv.* in Holland. 1676. Dutch loom *intro.* 1738. Lewis Paul, Birmingham *inv.* rollery-spinning loom. 1769. Richard Arkwright revived

Lewis Paul's process. 1794. Cam-operated loom *inv.* by William Bell, of Milton, near Dumbarton. 1800. Joseph Marie Jacquard (Fr) (1752–1854) *inv.* pattern-weaving loom. 1807. First steam-driven loom in use. 1850. Jacquard loom first used in England. 1850. Barlow *imp.* Jacquard loom.

Loom, Power 1768. Power loom suggested by M. de Gennes (Fr). 1785. *Inv.* by Rev. Edmund Cartwright. 1789. Loom driven by steam-engine by Austin, of Glasgow. 1792. William Kelly, of Glasgow *des.* power loom for cotton weaving. 1803. Crank-driven power loom *inv.* Charles Todd, of Bolton.

Loom, Silk 1717. Jurine, a lacemaker of Lyons *inv.* loom. 1740. Vaucanson *inv.* loom. 1767. Jaubert and Rocamus, of Provence *inv.* loom. 1790. Claud Rivey (Fr) *inv.* loom. 1792. Jean Paulet (Fr) *inv.* loom. 1860. G. Bounelli, of Milan *inv.* electrical setting of Jacquard loom needles. 1864. Harrison *inv.* compressed-air loom.

LORENTZ, Hendrik Antoon (Hol) (1853–1928) 1875. *Prop.* mathematically based theory of reflection and refraction of light.

LORENZ, Adolf (1854–1923) *Dev.* formula for molecular refractive power.

LULL (LULLY), Raymond (*c.* 1232–1316) Made earliest attempts to symbolize chemicals and chemical reactions.

Lubricants and Lubrication *c.* 1400 B.C. Fatty matter used to lubricate chariot wheels. 1787. Oil axle-box *inv.* John Collinge. 1827. Ring-oiler fitted to a steam locomotive journal bearing. 1831. Samuel Hall *pat.* pressurized oil lubricating system for steam-engines, by "the injection of a uniform stream of oil by a force-pump, using the same oil over and over again. 1883. Pressurized lubricating system devised by B. Tower. 1889. G. E. Bellis *inv.* forced feed lubrication system for the Bellis and Morcom high-speed steam-engine. 1890. Pressurized lubrication system *inv.* by Albert Pain. 1893. Mineral oils first used for high temperature lubrication. 1909. "Castrol" oil *intro.* by Charles C. Wakefield. 1916. "Tecalemit" high-pressure grease-gun *inv.* by F. D. Stone and O. U. Zerk. 1922. "Enots" (Messrs. Stone, Ltd.) grease-gun *inv.* by Stone and Benton. 1935. Thio-ether *disc.* as an additive to lubricants for hypoid gears. (*See also* Nylon bearings.)

LUMIÈRE, Louis (Fr) 1896. *Inv.* cinematograph (*q.v.*).

Luminescence 1845. Luminescence of quinine sulphate *disc.* Sir J. F. W. Herschel. (*See also* Electro-luminescence.)

LUMMER, Otto (Ger) (1860–1925) 1889. With Eugen Brodhulm *imp.* and marketed the simple photometers *inv.* by Robert Wilhelm Bunsen and Benjamin Thompson (Count Rumford).

Lutecium (element) 1907. *Disc.* by Georges Urbain in original "element" ytterbium (*q.v.*).

Lyddite 1888. Explosive first used at the artillery range at Lydd, Sussex. (*See also* Explosives and Picric acid.)

Lyons Blue (dye) 1859. First prepared by Nicholson.

Lymphatic System (anatomy) 1653. Existence first demonstrated by Thomas Bartholin, of Copenhagen.

MACADAM, John Loudon (1756–1836) Pioneered construction of roads from gravel, stone chips and loam. From 1815 the water-bound "macadam" road reigned supreme and was adopted by Napoleon for most of the continental "national" roads he built.

MACH, Ernst (Aus) (1838–1916) 1883. *Prop.* original concepts of mass, absolute space and absolute time, with particular regard to the so-called principle of economy. *Prop.* axiom of action and reaction and a philosophy of science.

Machine-gun 15th cent.

"Organ," or "ribandequin" guns used. *c.* 1860. Reyner gun ("rifle-battery") used during American Civil War. 1876. Montigny *inv.* "mitrailleuse" firing 150–200 shots per minute. Dr. Gatling (U.S.) *inv.* machine-gun with revolving barrels firing 400 shots per minute. Hotchkiss and Nordenfelt machine-gun *inv.* firing 600 shots per minute. *c.* 1881. Gardner machine-gun used in British Navy and in the Sudan, firing 300, 600, or 1,200 shots per minute. Maxim automatic machine-gun *inv.* Sir Hiram Maxim (U.S.). *See* Cannon.

Machine Tools *See* Gear-cutting, hobbing, milling and other machines; and the Lathe.

McCORMICK, Robert (U.S.) (1780–1846) Pioneered the use of steel in the *const.* of binding, mowing and threshing machinery for agriculture (*q.v.*).

Mackintosh (waterproof material) 1791. Fabric rubberized by Samuel Peal. 1822. Fabric double-shell rubberized by Charles Mackintosh. 1837. Fabric single-sheet rubberized by Thomas Hancock. (Brother of Walter of steam road coach fame.)

MACLEAR, Sir Thomas (1794–1879) *c.* 1840. With Bessel, Bradley, Struve and Henderson, pioneered work of determining the parallax of fixed stars.

McMILLAN, Kirkpatrick, of Thornhill, Dumfriesshire. 1839. *Const.* crank-operated hobbyhorse on which he rode the 70 miles from his home to Glasgow in June, 1842.

McNAUGHT, J. 1825–30. *Imp.* steam-engine indicator by lightening its parts and using a revolving recording cylinder. 1845. Pioneered the compound steam-engine by converting a simple-expansion engine to compound working and *pat.* the method used.

MACQUER, Peter Joseph (1718–84) *c.* 1782. With Berthollet founded modern technique of dyeing by *disc.* true action of mordants and air.

Madder 1666. Colbert established crop for dyeing purposes at Avignon, France.

MÄDLER, Johann Heinrich von (Ger) (1794–1874) 1834–36. *Pub.* with Wilhelm Beer (Ger) four-volume topographical charts of the moon.

Magenta (*See* Dyes, synthetic.)

Magic Lantern (projector) *Inv.* by della Porta (1545–1615). *Imp.* by Athanasius Kircher (Ger) (1602–80), who used lenses in a primitive magic lantern. 1650. Christian Huygens cast pictures from printed glass on to a screen.

Magnesium (metal) 1808. First prepared by Sir Humphry Davy. 1830. First commercially *prod.* in Germany, and properties first made known by Bussy.

Magnetism, Animal 1774. First *intro.* at Vienna by Jesuit Father Hehl. *See also* Mesmerism.

Magneto *See* Ignition, motor-car engine.

Magnetometer *Inv.* Gauss (1777–1855).

Magnus Effect *See* Ship, rotating-cylinder.

Malic Acid 1785. *Disc.* Carl Wilhelm Scheele (Ger) (1742–86).

MALIGRANI, Arturo (It) 1894. *Inv.* phosphorus "gettering" process for assisting to create vacuum in electric lamp bulbs.

MALPIGHI, Marcello (It) (1628–94), of Bologna. Biologist and embryologist. First to see and sketch leaf stomata and to write first monograph on an invertebrate animal—the silkworm. 1673. First to *obs. dev.* of embryo chick.

MALUS, Étienne-Louis (Fr) (1775–1812) 1808. *Disc.* the practical application of polarized light.

"Man-engine" (Passenger mine-lift) 1830s. Originated in Hartz Mountains mines. 1841. *Intro.* into Cornish tin and copper mines. 1920. Last man-engine closed at Laxey, Isle of Man.

Mallet, Mason's Tool *des.* originated in Ancient Egypt. MS. dated at 1370 B.C. and of precisely similar form to the modern tool have been found in Egypt.

Manganese (element) 1774. *Disc.* by C. W. Scheele (Ger) (1742–86). (According to some authorities it was *disc.* by Bergmann and *isol.* by Gahn in 1774.)

Maganese Steel 1888. *Inv.* by Robert Hadfield.

Manometer *See* Pressure-gauge.

MANSFIELD, Charles Blackford (1819–55) 1849. *Disc.* way of preparing benzene commercially from coal tar.

Maps c. 2400 B.C. First evidence of map-making for land taxation purposes by Sargon of

Akkad, Babylon. Inscription on baked clay was used. 568 B.C. Maps *"inv."* by Anaximander of Miletus. A.D. 1568. Gerard Kremer (Mercator) *intro.* the projection method known by his name today. 1815. William Smith (1769–1839) *prod.* first coloured geological map.

MARALDI, Giacomo Filippo (It) (1665–1729) One of the first astronomers to measure the period of rotation of the planet Mars.

MARCONI, Guglielmo (It) (1874–1937) *Disc.* how to transmit and receive wireless (radio) waves. *See also* Radiotelegraphy.

MARCUS, Antonius de Dominis (1566–1624) First physicist who tried to explain the optics of the rainbow.

Margarine 1869. Mege-Mouries (Fr) *pat.* in England process for making. 1910. Process of hardening margarine by addition of oleïne *disc.*

MARGRAAF, Andreas Sigismund (Ger) (1709–82) 1747. *Disc.* sugar in beetroot. 1758–59. *Disc.* method of preparing a fixed alkali from common salt.

MARIOTTE, Edmé (Fr) (1620–84) 1679. Independently of Boyle, *disc.* the law of relation between the volume and pressure of air in an enclosed vessel. (The law now known as Boyle's Law (*q.v.*).)

Mars (planet) Period of daily rotation measured by Cassini, Maraldi and Huygens.

MARTIN, Archer John Porter (*b.* 1810) With Le Chatelier *inv.* a reverberatory furnace for steel-making which was later lined with high-class silicious bricks to produce, in 1873 Siemens-Martin open-hearth mild steel. *See* Steel.

MARUM, Martinus van (Hol) (1750–1837) 1787. With A. Paets van Troostwijk (1752–1837), succeeded in liquefying ammonia.

MASER (Acronym = Microwave Amplification by Stimulated Emission of Radiation) 1917. Concept of stimulated emission of radiation *intro.* by Albert Einstein. This concept was later *dev.* by Dirac, who predicted the properties of stimulated emission. 1952. Amplification of microwaves first suggested by J. Weber. 1955. Microwave amplification demonstrated by C. H. Townes, using ammonia. 1956. Solid-state ruby amplification suggested by N. Bloemberger. 1958. Extension of maser techniques to visible light suggested by A. L. Schawlow and C. H. Townes. *See* Laser.

MASKELYNE, Nevil (1732–1811) 1774. Deduced the average density of the earth as 4.71 by measuring the deviation of a plumbline from the vertical at Mount Schiehallien, Scotland.

Mass-production 1794. Continuous fully mechanized flour-mill *des.* and built by Oliver Evans (U.S.). 1833. Ships' biscuit-making mechanized for the British Royal Navy. 1869. First "dissembly"-line *des.* for the dismembering of pigs at Chicago; together with a continuous monorail for pork packing. The forerunner of modern assembly-line. 1898. Eli Whitney (U.S.) first made mass-

production possible by manufacturing components to sufficiently accurate limits for interchangeability. 1903. First modern mass-production plant at mail-order firm of Sears, Roebuck, Chicago. *c.* 1910. System *intro.* by Henry Ford for production of his "Model T" motor-car. 1923. First transfer machine *intro.* into Morris Motor Works. 1924. James Archdale *des.* his first auto-transfer machine.

Mastoiditis (surgery) First operation for mastoiditis made by Jean Louis Petit (1674–1750), of Paris.

Mast, Ship's 1240 B.C. *Inv. att.* to Daedalus, of Athens.

Matches A.D. 970. Thao-Ku records Chinese "light-bringing slave," known as "yin kuang nu"; which later became the commercial "fire-inch-stick," or Huo-Tshun, as sold in the markets of Hanchow. 1270. This *inv.* mentioned by Marco Polo. 1530. Sulphur matches first mentioned in England. 1680. Robert Boyle used sulphur-tipped splints drawn through phosphorus-impregnated paper. *c.* 1745. Flint and steel wheels for miners *inv.* Charles Spedding, of Whitehaven, Cumberland. 1780. Phosphorus matches in use in England. 1780. "Phosphoric Candle" of phosphorus-impregnated paper in glass tube, *intro.* into France. 1805. Matches tipped with potassium chlorate and sugar, dipped in sulphuric acid to ignite. (*Inv.* by Chanel.) 1827. First modern matches sold by John Walker, of Stockton-on-Tees. 1830. J. F. Kammerer (Ger) *inv.* matches using yellow phosphorus, sulphur and potassium chlorate. 1833. Charles Sauria (Fr) *pat.* phosphorus matches. 1834. Matches finally superseded flint-and-steel. 1842. Matches first machine-made. 1851. Non-poisonous matches of amorphous phosphorus made by Arthus Albright, of Birmingham after acquiring the *pat.* to produce amorphous phosphorus already taken out in 1845 in Austria by Anton Schröetter, of Vienna. 1852. J. E. Lundström and his brother (Swed), of Jonköping, made safety matches with amorphous phosphorus on the box side. (According to von Meyer's *History of Technical Chemistry*, phosphorus matches were also *intro.* by Irinyi, of Pesth, Römer, of Vienna, and Moldenhauser, of Darmstadt.) 1860. Wax match-making machine *inv.* Louis F. Perrier, of Marseilles.

MAUDSLAY, Henry (1711–1832) Prolific inventor responsible for *inv.* of screw-cutting lathe, two-point chuck and lathe slide-rest.

MAUPERTUIS, Pierre-Louis Moreau de (Fr) (1698–1759) 1744. *Prop.* the principle of least action, as applied to pure mechanics.

MAUROLYCO, Francesco (It) (1494–1575) Great optical mathematician whose work *Photismi de Lumine* was not published until 36 years after his death. (Preceded Kepler in many of his ray refraction theories.)

MAXIM, Sir Hiram S. (1840–1916) Prolific inventor responsible for *inv.* the machine-gun bearing his name, together with many model and full-sized heavier-than-air flying machines.

MAXWELL, James Clerk (1831–79) 1868. *Inv.* automatic control system. 1873. *Dev.* the Faraday–Maxwell theory of electricity.

MAYER, Julius Robert von (Ger) (1814–78) *Prop.* as a "supreme and universal law of nature," his so-called "principle of energy," which claimed that the total amount of energy in the universe is constant.

MAYR (Marius) Simon (1570–1624) Claimed that he had *disc.* the moons of planet Jupiter in 1609 (before Galileo).

Meat Extract 1838. Baron von Liebig (Ger) (1803–73) reflesh food to 15 per cent of its bulk (weight) by dessication.

Megaphone *c.* 1666. *Inv.* Sir Samuel Morland (1625–95).

MEIKLE, Andrew (1719–1811) 1788. *Inv.* first useful threshing-machine, which threshed the corn, blew away the chaff, and separated grain and weeds by sieving. (Also various windmill *invs.*)

MENDELÉEV, Dimitri Ivanovich (Rus) (1834–1907) 1872. Proved the Periodic Law of Atomic Weights of the Elements.

Mendelevium (element) 1954. *Disc.* by Glen Seaborg (U.S.).

Mendelism (genetics) 1865. Abbé Mendel *pub.* his work. 1900. Mendel's work re-*disc.* and made known to world.

Mercerization (textiles) *Inv.* John Mercer (1791–1866), of Blackburn, Lancs.

Mercury (metal) Pre-750 B.C. Mercury *disc.* 1540. First used in silver refining. 1640. First used in refining gold.

Mercury, First Transit of Planet 1631 (Nov. 7). First time *obs.* by Gassendri (1592–1658).

Mercury Fulminate 1700. Impure first made by Johann Kunckel (Ger) (1630–1703). 1800. Pure first made by E. C. Howard. 1807. Applied to small arms ignition by Rev. Forsyth, of Aberdeen. (*See also* Percussion cap.)

Meridian, Measurement of Arc of 1669–70. First measured by Jean Picard (Fr), north of Paris. 1735. Degree measured by Charles-Marie de la Condamine (Fr) (1701–74) and Pierre Bouguer (Fr) (1685–1719), in expedition to Finland and Peru; when earth shape was established as an oblate spheroid.

Mesmerism 1766. *Intro.* by F. A. Mesmer (1734–1815).

Mesons, "K" (θ and τ) 1953. Anomaly between θ and τ particles *disc.* in England by R. Dalitz.

Mesons (Yukawa particles) 1936. π mesons first detected in upper atmosphere by Carl D. Anderson. 1947. μ mesons *disc.* by Powell, C. M. G. Lattes (Braz), and G. P. S. Occhialini (It).

Metals, High-temperature "creep" of 1910. First *obs.* by Andrade. 1935. R. W. Bailey used Andrade's formulae to withstand high temperatures and pressures.

Metals, Granulation of Decoration made by soldering droplets of granulated metal to a base found in Tutankhamen's tomb. 1782. Watts, a plumber of Bristol formed lead shot by pouring the molten metal from the tower of St. Mary Redcliffe church into water below.

Metallic Packing 1797. *Inv.* Rev. Edmond Cartwright (1743–1823), and the same year independently by Barton.

METCALF, John (1717–1810) One of the pioneers of road *construction* in England. (*See also* Macadam, J. L.)

Meteorology 1823. First systematized by Prof. Daniels. (*See also* Anemometer, Barometer, Thermometer, etc.)

Methyl Violet (dye) 1867. First *prep.* by August Wilhelm von Hoffman (Ger) (1818–92).

Metonic Cycle (relation between lunar and solar calendar) 432 B.C. *Obs.* and proposed by Meton, of Athens. 330 B.C. Cycle corrected by Callippus (*b.* 370 B.C.).

Metronome *Inv.* by Maelzel (1772–1835).

MEYER, Julius Lothar (Ger) (1830–95) 1869. First plotted atomic volumes against atomic weights and thereby proved that the properties of the elements are a periodic function of their atomic weight. (*See* Elements, Periodic tables of.)

Mezzotint (engraving) 1643. Method *inv.* Col. de Siegen.

MICHAUX, Ernest (Fr), of Paris 1865. With his workman Pierre Lallement, made a crank-driven velocipede. There is a monument to him at Bar-le-Duc. (*See* Bicycle.)

MICHELSON, Albert

Abraham (1852–1931) *c.* 1881. Pioneered measurement of the speed of light.

Micrometer 1638. *Inv.* by English astronomer William Gascoigne. 1640. Spider-thread micrometer *inv.* by Gascoigne. 1667. Screw adjustment added to Gascoigne's micrometer by Adrien Auzout (Fr). 1678. Ole Romer (Den) (1644–1710) *inv.* Double-image micrometer. 1740. Circular micrometer suggested by Boscovitch. 1742 Boscovitch's micrometer adopted by astronomer Lacaille (Fr). 1777. Dioptric micrometer *inv.* by Jesse Ramsden (1735–1800). 1792. Engineer Jean Richer (Fr) (1630–96) *inv.* micrometer which could divide a "ligne" ($\frac{1}{12}$ in. into 1,200 parts. 1848. Jean-Laurent Palmer (Fr) *imp.* Gascoigne-type micrometer and *prod.* first practical micrometer which could measure to 0.05 millimetre. Henry Maudslay (1711–1832) made micrometer to measure to $\frac{1}{10000}$ in. 1856. Whitworth made micrometer accurate to $\frac{1}{10000}$ in. 1869. Micrometer *imp.* by J. R. Brown and Lucien Sharpe (U.S.). 1875. Bar micrometer *inv.* by Brown and Sharpe. 1887. Screw-thread micrometer *inv.*

Microphone 1837. Prof. Charles G. Page (U.S.) *disc.* that magnetizing an iron bar electrically produced therein a "click." 1854. Charles Bourseul suggested use of flexible disc on which sound would cause a battery circuit to make and break, and so produce vibrations on another disc. He made experiments, but proceeded no further. 1856. Du Moncel *disc.* that increased pressure on an electrical contact diminished its resistance. 1860. Philip Reiss, of Freidrichsdorf *inv.* a microphone for music. 1876. Alexander Graham-Bell (U.S.-Scot) exhibited his microphone at Philadephia, and Elisha Gray independently *pat.* a version of the same instrument. 1877. Emil Berliner and T. A. Edison both *inv.* contact microphones. Moving-coil (dynamic) microphone independently *inv.* by Charles Cuttris (U.S.). and E. W. Siemens (Ger). T. A. Edison *inv.* carbon microphone. 1878. David E. Hughes *imp.* on Moncel's idea and was first to use word "microphone." 1878. Henry Hunnings *inv.* carbon-granule microphone. 1880. A. E. Dolbear *inv.* condenser microphone. 1890. Anthony C. White *inv.* solid-back, button microphone. 1916. Dolbear's *inv. dev.* practically by E. C. White. 1919. A. M. Nicholson first to make crystal microphone, using Rochelle salt. (*See* Diezo-electricity.) 1923. Ribbon microphone *inv.* by W. H. Schottky and Erwin Gerlach (Ger). 1931. Moving-coil microphone brought to perfection by E. C. Wente and A. L. Thomas. Ribbon microphone perfected by H. F. Olsen. 1932. Crystal microphone brought to perfection by C. B. Sawyer. 1935. Condenser microphone *dev.* by H. J. von Braunmühl and W. Weber (Ger). (*See also* Telephone.)

Milky Way 1761. J. H. Lambert (Ger) *prop.* first theory of the Milky Way.

Mill, Hand (*See* Quorn.)

Mill, Animal-powered 5th cent. B.C. Donkey-driven edge-runner mills or "Trapetum" for grinding corn made and used at Athens (300 B.C.). Also used for crushing olives, and in the silver mines at Lorion. (Generally used in Mediterranean countries.)

Mill, iron-rolling 1798. Continuous iron-rolling mill *inv.* by William Hazeldine. 1856. Three-high iron-rolling mill *inv.* by Christopher Polheim (Swed) for rolling sections. Used first at Motala. 1862. Re-*inv.* by George Bedson, of Manchester. 1862. Three-high mill in use in Birmingham.

Mill, Slitting (for hand-made nails) 1588. Imported from Flanders and established at Dartford, Kent, by Bevis Bulmer. 1628. *Intro.* Richard Foley (1580–1657) at Hyde, near Stourbridge.

Mill, Sugar 1449. Three-roller (wooden) sugar mill *inv.* Pietro Speciale, Prefect of Sicily; suitable for water or ox drive. 1653. Iron-clad roller mill *intro.* by George Sitwell. 1754. John Smeaton *inv.* triangular sugar mill with three rollers. 1773. Dumb turner John Fleming *pat.* the "Wallerer Wheel." 1802. Richard Trevithick and Vivian (Cornwall) *pat.* steam-driven roller sugar mill. 1805. Steam "Cane-engine" built by James Cook, of Glasgow. *c.* 1830. Matthew Boulton and James Watt design sugar mill. 1840. Roller sugar mill *pat.* by James Robinson. 1858. Hydraulic pressure *intro.* in sugar-milling by Jeremiah Howard.

Mill, Tide 1170. Earliest recorded in Britain at Woodbridge, Suffolk.

Mill, Water 398 B.C. Water mills protected by Roman edict. 5th cent. B.C. Overshot waterwheel in use in Athens, below the Parthenon. *c.* 85 B.C. Antipater of Thesalonica mentions "Shetland-type" water mill. Late 1st cent. B.C. Vitruvius Pollio, Marcus (*c.* 50–26 B.C.) *desc.* mill with undershot waterwheel geared to millstone. (No evidence of its use.) Roman General Belisarius said to have *inv.* the floating, current-driven mill. A.D. 2nd cent. Undershot water-wheels evinced by archaeological finds alongside the Roman Wall, England. *c.* 315. Large water mill using water-wheels in series, built at Barbegal, France, to grind corn for 50,000 persons. 3rd cent. "Shetland" (Norse) horizontal water mill in use in Ireland. 536. Belisarius devised water mill to float on the River Tiber ("Current-wheels"). 762. Earliest documentary evidence of corn-grinding water mill in England. 1086. Domesday Book records 5,264 mills (mostly, probably water) in England. 12th cent. Undershot water-wheels *illus.* in French MSS. Floating water mill in use on River Seine, at Paris. (*See also* Water-wheel.)

Mill, Wind 7th cent. Horizontal wind-mill probably known. Early 10th cent. Horizontal wind-mill used in Arab countries. 10th–11th cents. Horizontal wind-mill in use in Low Countries. *c.* 1180. Post-type wind-mill used in Normandy. 1191. Earliest reference to wind-mill in England at Bury St. Edmunds.

Mills (general machinery of) 1588. Ramelli *desc.* new roller mill. 1502. Boller mechanized the bolting process 1588. Ramelli and Veranzio mechanize bolting-mill. 1637. First English *pat.* for mill taken out by George Manby and Thomas Lidell. (24 more *pats.* in the subsequent 162 years.)

Milling-machine 1848. *Desc.* by Eli Whitney (U.S.) and built for sale. Howe made first milling machine. 1861. Universal milling-machine *inv.* by J. R. Brown and Lucien Sharpe (U.S.) for cutting drill-spirals. This was the first milling machine. It was suggested to Brown and Sharpe by Howe. 1862. W. B. Bement (U.S.) *inv.* vertical milling-machine. 1864. J. R. Brown *inv.* formed cutter for milling-machine. 1870. James Watson (U.S.) *pat.* milling-machine. 1883. Swing-spindle and movable table *inv.* L. Cosgrove. 1885. Swivelling spindle milling-machine *inv.* Joseph Saget (Fr). George Richards *inv.* milling-machine. 1900. Milling-machine first *intro.* into England.

MILLIKAN, Robert Andrews (U.S.) (1868–1963) Physicist who first accurately determined the charge of an electron.

Mimeograph 1870. *Inv.* Thomas Alva Edison (U.S.) (1847–1931).

Mineralogy 1695. First made an accurate science by John Woodward.

Mine-haulage 1812. George Stephenson adapted a pumping engine to. 1844. Endless-rope steam-operated system *intro.* John Buddle, at Wallsend.

Mineral Waters (artificial) 1821. First made by Dr. Struve, of Dresden. 1825. Flavourings first added. (*See also* Gases.)

Minuet Waltz 1889. *Inv.* by Johann Strauss.

Mirror 1250 B.C. First mentioned, Job xxxvii. 18. (*c.* 287–212 B.C. multiple faceted concentric mirror *inv.* by Archimedes of Syracuse.) A.D. 1300. First made from glass at Venice. 1673. First used in England. (*See also* Burning-glass.)

MITCHELL, L. E. (U.S.) 1919. With A. J. White (U.S.) re-*inv.* "new" pipeless electric-light bulb. (This had already been *inv.* by Jaeger (Ger) in 1903.)

MITCHELL, John (1729–1823) 1750. First to *prop.* the inverse square law of magnetic attraction and repulsion, which was in 1785 proved with a torsion balance by Charles Augustin de Coulomb (Fr) (1736–1806).

Mnemonics 477 B.C. Science *inv.* by Simonides the Younger.

MÖBIUS, August Ferdinand (Ger) (1790–1868) With Heinrich Günther Grassman (Ger) founded the so-called direct methods of calculation now used in vector analysis.

Modelling, Wax 328 B.C. Bust first modelled in wax.

"Moho" (**"Mohorovičić"**) Discontinuity of the earth's crust first recognized by seismologist Mohorovičić (Jugo-Slav).

MOISSAN, Henri (Fr) (1852–1907) 1892. Evolved the means

of increasing efficiency of the electric arc metallurgical furnace.

Molecule 1865. Diameter of M. first *est.* by Loshmidt; by Stoney, 1868; by Sir W. Thomson, 1870.

Molybdenum (element) 1778. *Disc.* Torberu Bergmann, with C. W. Scheele (1742–86). 1782. *Isol.* by Hjelm.

Molybdic Acid. 1778. *Disc.* by C. W. Scheele.

Momentum (moment) 1745. Principle of areas, or principle of moment of momentum *prop.* by Daniel Bernoulli (1700–82).

MONDINO of Bologna (*c.* 1270–1326) The first practical anatomist.

Monel-metal (nickel-cobalt-iron alloy) 1905. *Disc.* by Ambrose Monell, president of International Nickel Corp., U.S.A.

Monorail 1821. First monorail *inv.* by Henry Robinson Palmer. Others *inv.* by: 1825, Henry Sergeant (U.S.); Jacob Jedder Fisher; 1829, D. Maxwell; 1830, J. Stimpson (U.S.); 1831, Bryant and Gyatt (U.S.); 1832, J. Richards (U.S.); 1837, U. Emmons (U.S.); 1845, William Newton; 1846, Sir Samuel Brown; 1872, E. Crewe (U.S.); 1876, Gen. Leroy Stone (.U.S.); and 1884, C. F. M. T. Lartigue (Fr), who erected the passenger and freight monorail—the Listowel and Ballybunion Railway in Ireland.

Monotype Casting Machine 1841. Type composing machine *inv.* M. Ballanche, of Lyons. 1887. "Monotype" machine *inv.* Tolbert Lanston, clerk at Pension Office, Washington,

D.C., U.S. Dr. Mackie, of the *Warrington Guardian inv.* and *dev.* a steam-operated composing machine.

MONTGOLFIER, Joseph (1740–1810) and **Étienne** (1745–1799) of Annonay, France. 1783 (June 5). First balloon (hot-air) ascent without passengers at Annonay. Aug. 27. Second ascent from Champ de Mars, Paris. (*See* Balloon.)

Moon, Maps of the 1645. Chart *prod.* by Michael Florent van Langeren (Hol) (*d.* 1675). 1647. Chart *prod.* by Johannes Hevelius (Hevel, or Hewelke) (Hol) (1611–87). 1834–36. J. H. von Mädler, with W. Beer, *pub.* four-vol. topographical charts of Moon.

Moon, Libration of the 1637. *Disc.* by Galileo (1564–1642).

Mordants Early 17th cent. Cornelius Drebbel (Hol) *disc.* use of tin salts as mordants. *c.* 1782. True action of mordants *disc.* P. J. Macquer (1718–84).

MORLAND, Sir Samuel (1625–95) 1674. *Inv.* efficient packing rings for steam-engine pistons.

Morphine (morphia) 1803. First *isol.* from opium by Charles Derosne (Fr) (1780–1846). 1806. *Isol.*, independently from same source by Sertürner.

MORSE, Samuel Finley Breeze (U.S.) (1791–1872) 1840. *Inv.* the telegraphic code bearing his name. 1844. Sent first electric telegraph message from Washington, D.C. to Baltimore.

MORT, Thomas Sutcliffe (Australia) (1816–78) 1861.

Built first machine-chilled cold store at Sydney, N.S.W.

Mortar *See* Cement.

MORVEAU, Louis-Bernard, Baron Guyton de (Fr) (1737–1816) 1789. Proposed tentative chemical terminology in his book *Traite élémentaire de chemie*.

MOSELEY, H. G. S. (1887–1915) *Disc.* that atomic nucleus had an electric charge the size of which is characteristic of the atom; the numerical value of this charge being known as its atomic number.

Motor-bus 1869. Three-wheel steam motor-bus *des.* Andrew Nairn of Leith (Scotland). 1889 (Feb.). W. C. Bersey built for Michael Radcliffe-Ward a 3½ ton, battery-driven motor-bus. Tried in London between Victoria and Charing Cross railway stations. 1896. Converted horse-bus ran in London. 1896. Daimler-built petrol lorries in use. 1897. First motor-bus to run in London between Notting Hill and Marble Arch. Twin-cylinder, 17-seats; 9 miles per gallon. Sponsored by H. J. Lawson. 1899 (June 15). First motor-bus service in world inaugurated between Künzelsau and Mergenthein, Germany. Fischer *inv.* petrol-electric motor-bus. 1901. Daimler 11.8 h.p. wagonettes run in service on three routes in London: Piccadilly-Putney, Oxford Circus–Kilburn, and Clapham Junction–Streatham. 1902. 27-seat Cannstadt-Daimler motor-bus ran service between Eltham and Lewisham, London. 1902–9. Steam motor-buses tried in London. 1906. Petrol-electric motor-bus

inv. W. A. Stevens, of Maidstone. 1931–39. Henschel and Son, of Cassel manufactured steam Motor-buses under Doble *pats.* (*See also* Road vehicles, steam.)

Motor-car, Electric 1891. William Morrison, of Des Moines, U.S., first to attempt to run motor-car from accumulators.

Motor-car 1820. Rev. Edward Cecil *des.* an external-combustion-engined vehicle. 1823. Samuel Brown *des.* an external-combustion-engined vehicle (and pumping-engine). *See also* Engine, internal-combustion. 1860–63. Étienne Lenoir (Fr) (1822–1900) drove a gas-engine-propelled road vehicle. 1875. S. Marcus drove a road vehicle fitted with a gas-engine. *c.* 1884. Karl Benz independently *des.* a four-cycle petrol engine and tried it in a road vehicle (*pat.* 1886). *c.* 1885. Gottleib Daimler (1834–90) *inv.* high-speed petrol engine and tried it in several road vehicles. 1890. Benz *prod.* first four-wheeled motor-car. 1891. Charles E. Duryea (U.S.) makes first motor-car in U.S. 1891–92. J. D. Rootes "Petrocar" features *pat.* 1892. Rootes' motor tricycle built. 1893. Benz *intro.* Ackermann steering on his motor-cars. 1894. Panhard-Levassor *dev.* Daimler-engined motor-car in France. 1895. Benz makes first racing-car and delivery-van. Rootes and Venables three-wheeled "Petrocar" built. *See also* Motor-cycle. 1895 (June). First motor-car— a Canstadt-Daimler, brought to England by J. A. Koosen.

1896. Peugeot *pat.* four-stroke, twin-cylinder petrol engine and marketed it the following year in a tubular-chassised motor-car. Frederick W. Lanchester *des.* and builds his first unique motor-car. 1897. Stanley, of Newton, Mass., U.S., *intro.* his first steam motor-car. 1901. First Mercédes motor-car built by Daimler. 1902. First rear-engined motor-car a U.S. Cadillac, appears. 1904. M. Pope (U.S.) *inv.* exhaust gas-heated muffle. 1909. Darracq *intro.* pressed-steel motor-car chassis as unit with bodywork. (There is, apparently, no evidence of actual *prod.* by the Spanish engineer who *des.* it.) 1931. Pressed-steel motor-car bodies first *prod. See also* individual technical details and individual motor-car accessories.

Motor-cycle (and **Tricycle**) *c.* 1886. Gottleib Daimler (Ger) (1834–90) makes the first motor-cycle with $\frac{1}{2}$ h.p. petrol engine. 1892. J. D. Rootes *des.* and made single-cylinder two-stroke motor-cycle. 1894 Hildebrand and Wolfmüller (Ger) *pat.* and market four-cycle, twin-cylinder machine capable of 24 m.p.h. 1896–97. Col. H. Capel Holden *pat.* first four-cylinder motor-cycle. 1900. Clement-Gerrard detachable cycle-motor unit *intro.* 1901. "Motosacoche" detachable cycle-motor unit *intro.* by H. and A. Dufaux, of Geneva. 1902. Michael and Eugene Werner (Fr) build motor-cycle after having in 1896 built a light "motorcyclette" with a De Dion-Bouton type of engine and clutch drive through chain.

1903. J. A. Prestwich *prod.* the first English proprietary motor-cycle engine—the "J.A.P." He also pioneered the vee-twin engine in England. 1905. 14 h.p. Peugeot-engined motor-cycle attained a speed of 86 m.p.h. at Brighton. 1908. A "J.A.P."-engined motor-cycle reaches 90 m.p.h. at Brighton. 1884. Edward Butler *des.* and *pat.* light, oil-driven tricycle, shown at Stanley Cycle Show of that year. The machine was built the following year. 1887. Brown *pat imp.* model which was built in 1888. 1889. Brown built a four-cycle engined vehicle which was run on Kentish roads.

Motor-cycle (and **Tricycle**), **Steam** 1818. Steam motor-cycle said to have had its initial trial in the Luxembourg Gardens, Paris. 1869. French Michaux velocipede (*see* Cycle) fitted with a single-cylinder Perreaux steam-engine unit. *c.* 1869. S. H. Roper (U.S.) made steam-driven velocipede. 1877. Meek, of Newcastle upon Tyne *prod.* a steam-driven tricycle. Cheylesmore pedal-cycle installed with a steam unit by Sir Thomas Parkyns and A. H. Bateman. 1884. L. D. Copeland, of Philadelphia, U.S. became first commercial producer of a motor-cycle of any kind by fitting a steam-engine unit into an American "Star" ordinary bicycle. 1887. Count Albert De Dion (Fr) (1856–1946) *const.* steam motor-tricycle. 1888. Léon Serpollet (Fr) (1858–1907) *const.* steam motor-tricycle. *c.* 1913. Pearson and Cox hot-tube, flash-boilered steam motor-cycle *intro.* commercially by Frank Giffin Carter, of Croydon. Production ceased owing to the outbreak of World War One.

Motor, Electric 1829. Joseph Henry (U.S.) (1797–1878) built electric motor. 1832. William Sturgeon, having *inv.* and made an electro-magnet in 1825, made a crude magnetic engine or electric motor. 1838. Thomas Davenport (U.S.) (1802–51) made electric engines (electric motors) and used them for driving workshop tools and a printing-press. 1838. M. H. Jacobi (1801–75) drove a boat by a magnetic engine. (Zénobie-Théophile Gramme, having *intro.* the ring-armature in 1870, in 1873 gave first demonstration of power transmission from an electric generator, through wires, to an electric motor ($\frac{3}{4}$ mile) driving a pump at the Vienna Exhibition.) *c.* 1887. Alternating-current electric motor developed. 1887–88. Nikola Tesla (1857–1943) *pat.* two-phase alternators and induction motors; thus inaugurating polyphase electrical engineering. (*See also* Ship, electric; Locomotive, electric; Railway, electric and Dynamo.)

MOUFET, Thomas (1553–1604) London naturalist who, pioneered the study of insects in his book *Theatre of Insects. Pub.* (Latin), 1634, and in English, 1658.

Moulds (casting) 328 B.C. Said to have been *inv.* by Lysistratus to cast wax figures. 1466. Casts from human face in plaster first taken by Andrea Verrochi (It). (*See also* Casting.)

Mould-making Machine (sand) 1850. *Inv.* R. Jobson.

Mouth-organ 1821. *Inv.* by Buschmann, of Berlin.

Mowing-machine *See* Reaping-machines.

Mowing-machine, Lawn 1830. *Inv.* Edwin Budding, of Stroud, Glos., after machine used for cutting pile on cloth. 1870 Horse-drawn mowing-machine *intro.* 1893. Steam-driven mowing-machine *intro.*

Mucic Acid 1780. *Disc.* C. W. Scheele (Swed) (1742–86).

MUDGE, Thomas 1755. *Inv.* the "English lever" watch escapement mechanism.

Mule (textiles) 1779 (1774?). *Inv.* Samuel Crompton. 1825. Made self-acting by Richard Roberts, who *imp.* upon it by adding the "faller wire" mechanism.

Mullerian Ducts (anatomy) *Disc.* by Johannes Müller (1801–58), of Coblenz. (He was the tutor of Herman von Helmholtz.)

Multiplier, Thermo-electric *Inv.* Melloni (1798–1854).

Muntz-metal 1832. *Inv.* G. F. Muntz, a brassfounder of Birmingham, who sold it for the sheathing of ships' bottoms.

MURDOCK (MURDOCH), William (1754–1839) Prolific inventor. 1785. *Inv.* and built an experimental model steam tricycle (cylinders $\frac{3}{4}$ in. bore × 2 in. stroke). 1799. *Pat.* worm-driven cylinder-boring machine; one-piece steam cylinder; double-dee steam-engine slide-valve (in place of James Watt's four poppet valves); a $\frac{1}{2}$ h.p. rotary steam-engine; and many other important *invs*.

Muscles Power of muscles to contract *disc.* by Haller (1708–1777), a pupil of Boerhaave (1668–1738), founder of organic chemistry.

MUSHET, David (1772–1847) Inventor of the steel-making process bearing his name.

Music *c.* 920. Notation *inv.* by Hucbald (*d.* 930). 1024. Notation *imp.* by Guido d'Arezzo; and also by Franco. Bishop Ambrosius of Milan (340–397), and Pope Gregory I (544–604), also *imp.* notation. *c.* 1410. Dunstable *inv.* polyphony. 1482. Equal temperament system devised by Bartolo Rames and later championed by Simpa Stevin. 1511. Mean tone system *prop.* by organ-builder Arnolt Schlick (Ger). 1577. Schlick's system elaborated by Franciscus Salinas, Spanish musician. 1636. Mersenne de-

tected harmonics and determined absolute frequencies.

Musical-box (carillon à musique) 1780. *Inv.* (cylinder-type) by Louis Favre, of Geneva. 1885. Card-disc record musical-box *inv.* by Paul Lochman. 1886. Steel disc record musical-box *inv.* by Paul Lochman. 1889. 27 in. diameter metal-disc musical-box ("Symphonium") made in U.S. *c.* 1890. "Polyphon" steel-disc musical-box developed.

Musical Glasses (harmonicon) 1651. *Inv.* in Nuremberg. *Intro.* as a musical instrument into England by Christopher Willibald von Gluck (1716–1787).

MUSSCHENBROEK, Petrus van (Hol) (1692–1761) With 's Gravesande (Hol), was the first to *des.* and demonstrate machines for measuring tensile, breaking, and bending strength of various materials used as beams, arches, etc.

N

Nails Used in Ur of the Chaldees to fasten together sheet metal. "Iron in abundance, for nails" mentioned in 1 Chronicles xxii. 3. Pre-1500. Made by hand by drawing small pieces of metal through a succession of graded holes in a

metal plate. 1741. Manufacture by smiths recorded in Walsall, England, where 60,000 persons employed in trade.

Nail-making Machine 1786. First *inv.* by Ezekiel Reed (U.S.). 1790. *Inv.* Thomas Clifford. 1851. Wire-nail-making machine made by Adolph Felix Browne, of New York City.

NAPIER (NEPER), John (Scot) (1550–1617) *Inv.* the calculating "bones" bearing his name, and logarithms.

Narcotine 1817. *Isol.* from opium by Robiquet (Fr). 1911. Synthesized by Perkin and Robinson.

NASMYTH, James (Scot) (1808–90) 1827. *Inv.* a steam road carriage. 1838. *Inv.* a steam hammer.

Navvy, Steam 1877. *Inv.* H. W. Ball. 1878. *Inv.* Messrs. Ruston, Proctor and Co., of Lincoln.

Naturalism (art) *c.* 1450. *Disc.* by Leone Battista Alberta and Piero della Francesca (It); artists.

Needle (sewing) 10,000–5,000 B.C. (Middle Stone Age). Bone needles 1–2 in. long with eyes bored from both sides found in cave-dwellings. Pliny the Elder (A.D. 23–79) mentions needles of bronze. 1370. Hook-eyed iron needles made in Nuremberg. 15th cent. Eyed needles made in Holland. 1545. Fine steel "Spanish" needles made by an Indian in Cheapside, London. The art then lost. 1650. Art of needle-making re-*disc.* in England by Christopher Greening, of Long Crendon, Buckinghamshire. He migrated to Redditch, Studley, and Alcester. 1860. Flap-needle *inv.* William Pidding, of London.

Needle, Sewing-machine 1907. Self-threading sewing-machine needle *inv.* Frank Giffin Carter, of Brighton, Sussex.

Needle (magnetic dip) 1544.

Disc. by German Pastor George Hartmann. 1576. *Disc.* generally *att.* to Wapping (London) compass-maker; Hartmann's prior *disc.* not being made known.

NEILSON, James Beaumont (Scot) (1792–1865) 1829. *Inv.* pre-heated air-blast for smelting; being followed by Cowper, in 1860, who *des.* the modern high towered blast-furnace.

Neodymium (element) 1885. *Disc.* Carl Auer von Welsbach (Ger) (1858–1929).

Neomycin (drug) 1948. *Disc.* Dr. Waksman. (*See also* Streptomycin.)

Neon (element) 1898. *Disc.* by Sir William Ramsay (1852–1916). (*See also* Argon, Helium, Krypton and Xenon.)

Neon Tube Sign 1909. Collie *obs.* that a bubble of neon gas in evacuated chamber of a Töpler pump acquired a red luminosity under electric discharge. 1910. Georges Claude (Fr) *intro.* neon tube sign.

Neoprene (artificial rubber) 1931. *Prod.* by Julius Arthur Nieuland (U.S.) by polymerizing chloroprene obtained from monovinyl-acetylene.

Neo-Salvarsan 1912. *Disc.* Paul Erlich (Ger). (*See also* Salvarsan.)

Nephelometer 1903. *Inv.* Theodore William Richards (U.S.).

Nephoscope (cloud speed-measuring instrument 1868. *Inv.* Karl Braun (Ger).

Neptune (planet) 1846. *Disc.* by Johanne G. Galle, at Berlin as a result of independent calculations by J. C. Adams (1819–92) and J. J. U. Leverrier (Fr).

Neptunium (element) 1940. *Disc.* McMillan and P. R. Abelson (U.S.). Isotope Np 237 *disc.* A. C. Wahl and Glen T. Seaborg (U.S.).

NERNST, Walther (Ger) (1864–1941) 1916. With Haber and Bosch synthesized ammonia from highly-compressed hydrogen and nitrogen. (First factory opened, 1917.)

Nerves (anatomy) 1814. Main types of nerves *disc.* by Sir Charles Bell (1744–1842).

Nerve-action, Reflex *Disc.* by Marshall Hall (1790–1857), of Nottingham. (Theory accepted only on the Continent.)

Net-weaving Machine 1830. Net-weaving machine to produce variable-pitch mesh *inv.* Alexander Buchanan, of Paisley.

NEUMANN, Franz Ernst (Ger) (1798–1895) 1831. *Disc.* that product of molecular weight and specific heat is approximately equal to the sum of the atomic heats of the constituent elements comprising a substance.

Neutron 1932 (Feb. 27). Presence announced by J. Chadwick (1891–).

Neutrino Particle 1956. Detected by F. Reines and C. L. Cowan at Los Alamos, U.S. (*See also* Feynman Theory.)

NEWCOMEN, Thomas (1663–1729) Blacksmith, of Dartmouth, England. From 1705 onwards made many *invs.* to *imp.* the atmospheric steam-engine.

NEWTON, Sir Isaac (1642–1727) Born- Woolsthorpe, near Grantham, England. Founded mechanics as an independent science, and applied it to nature. Established synthesis of terrestial and celestial mechanics by relating mechanics to theoretical astronomy. Made major *invs.* and *disc.* in optics.

Nickel 1751. *Isol.* by Cronstedt. 1775. First refined by Torbern Bergmann (1735–84). 1804. First prepared pure. 1822. Chinese "pak-tong," or white copper analysed by Fyfe and found to be an alloy of nickel, zinc and copper. 1843. Böttger first deposited nickel electrolytically. 1870. Nickel plating commercially *dev.* 1878. Nickel first prepared malleable by Fleitmann. 1889. James Ridley *disc.* use of nickel in alloy steels. 1890s. Ludvig Mond *inv.* carbonyl extraction process.

Nichol Prism 1828. *Inv.* by William Nichol (Scot) (1768–1851).

Nicotine 1560. *Intro.* into Europe by Jean Nicot (1530–1600), French consul at Lisbon. Brought seeds to Catharine de Medici. 1828. Alkaloid *disc.* by Possett and Reimann. 1893. Structure of the alkaloid *est.* 1904. Synthesized by Marc-Auguste Pictet (Fr).

Nicotinic Acid 1937. First *isol.* (from liver) by Elvehjem.

NIÉPCE, Joseph Nicéphore (Fr) (1765–1833) 1814. Commenced experimenting to produce camera pictures by action of light and finally succeeded in *prod.* permanent pictures. Entered into partnership with L. J. M. Daguerre, who six years after Niépce's death (1839) improved the process and termed it "Daguerreotype."

Niépce also made a velocipede ("céléripede").

NIEUWLAND, Julius Arthur (U.S.) (1878–1936) 1931. *Prod.* "Neoprene" by polymerizing chloroprene obtained from monovinyl-acetylene by the aid of hydrochloric acid.

Niobium (element) 1801. *Disc.* by Charles Hatchett.

Nitric Acid *c.* A.D. 800. *Disc.* by Giaber (Geber, or Yeber) (Arab). 1150. First *prod.* from saltpetre and alum in Italy.

Nitrogen (gas) 1772. *Disc.* independently by Joseph Priestley (1733–1804) and Lewis Morris Rutherdorf (1816–92). 1903. Kristian Birkeland (1867–1917) and Dr. Samuel Eyde (Nor) (1866–1940) fixed nitrogen gas electrolytically. (*See also* Nernst.) 1911. Active nitrogen *disc.* Lord Rayleigh (1842–1919). (Nitrogen cycle (botany): *see* Jean Baptiste Boussinggault (Fr).)

Nitro-cellulose 1830s. Pelouze and Bracconot (Fr) *prod.* nitrocellulose. 1846. Schonbein (Swit) made first nitro-cellulose explosive—gun-cotton (*q.v.*). (Nitro-cellulose *disc.* (?) 1841 by Schönbein.)

Nitrogen-cycle (botany) 1850. *Disc.* J. B. Boussinggault.

Nitro-glycerine 1846. *Disc.* Aloysius Sobrero (1812–88) whilst working in Pelouze's laboratory.

Nitro-methane 1872. First *disc.* as a nitro-derivative by Hermann Kölbe (Ger) (1818–84).

Nitrous Oxide (gas) 1772. *Disc.* Joseph Priestley (1733–1804). 1799. Anaesthetic effect of *disc.* by Sir Humphry Davy (1778–1829). 1823. Nitrous oxide first liquefied by Michael Faraday (1791–1867). 1824. Dr. Henry Hill Hickman, of Ludlow, England anaesthetized animals with nitrous oxide. (*See also* Anaesthetics.)

NOBEL, Alfred B. (Swed) (1833–96) 1867. *Inv.* dynamite, an explosive made by absorbing nitro-glycerine (*q.v.*) in the diatomaceous earth, kieselguhr.

Nobelium (element) 1954. *Disc.*

NOBILI, Leopoldo (It) (1784–1835) 1825. *Inv.* the astatic galvanometer (*q.v.*).

Nocturnal (survey instrument) 1235. *Inv.* by Raymond Lully (Lull, Lulle, or Lulli) (*c.* 1232–1316), of Palma, Majorca.

Notation, Literal (mathematics) Late 16th cent. *Inv.* Victa (Fr).

Notes (musical) 1681. Dr. Robert Hooke (1635–1703) calculated vibration of sounds by striking of teeth of revolving brass wheels. 1700. Joseph Sauveur (Fr) determined number of vibrations to a given musical note. *See also* Siren.

NOLLET, Jean-Antoine (Fr) (1700–70) 1746. Popular electrical experimenter who named the "Leyden Jar." (*See* Condenser, Electrical.)

Nuclear Fission 1938. *Disc.* by O. Hahn (1879–) and F. F. Strassmann (Germans); and named by O. R. Frisch (1904–).

Numbering-machine 1796. *Inv.* Joseph Bramah (1748–1814) for numbering Bank of England notes. 1850. Disc-type numbering and counting machines for table use *inv.*

Joseph John Baranowski, of London.

Numbers Early cents. A.D. Negative numbers first used in India. 1797. Geometrical interpretation of *prop.* by Carl Friedrich Gauss (1777–1855).

Numerals 9th cent. A.D. Ben Musa (Arab) brought into use Indian numerals later known as Arabian numerals.

Nut-milling Machine 1830. Self-acting nut-milling machine *inv.* James Nasmyth (1808–90).

Nylon 1938. Nylon (first wholly man-made fibre) announced by E. I. de Pont de Nemours (U.S.) as result of 10 years' research by W. H. Carrington (U.S.).

Nutation (of earth's axis) *Disc.* James Bradley (1692–1762).

OBEL, Matthias de l' (1538–1616) Naturalist. Made first attempt to classify plants by their structure.

Octobasse (music) 1849. Instrument *inv.* Jean-Baptiste Villaume (Fr).

Odometer 25 B.C. Mentioned by Marcus Vitruvius Pollio. 1724. *Inv.* by Meynier (Fr) and *imp.* by Hillerin de Boistissandeau (Fr) and attached to carriage wheel. 1756. Re-*inv.* by Hohlfeld (Ger).

Odontology 1839. Made a science by Prof. Owen.

ŒRSTED, Hans Christiaan (Den) (1777–1851) 1820. *Disc.* the magnetic field surrounding a wire carrying electric current.

OHM, George Simon (Dan) (1787–1854) 1826. *Disc.* and formulated the electrical law now bearing his name. 1843.

Made oustanding experiments with harmonic vibrations.

Oil *See* Petroleum and Lubricants.

Oil, Cod-liver Medicinal value of *disc.* by Dr. John Hughes Bennet (1812–1875), of Edinburgh.

Oil-on-water Experiment 1757. First made by Benjamin Franklin (U.S.) (1706–90).

Oilcloth 1754. First made by Nathan Taylor at Knightsbridge, London. (*See also* Kamptulicon.)

Oil Colours 1230. First used in painting in England. *c.* 1400. *Inv.* by Brothers Van Eyck (Hol). 1450. First used in painting in Venice by Veneziano.

Omnibus, Horse Blaise Pascal (1623–62) suggested. 1829. *Intro.* into London by George Shillibeer (July 4). 1854. Appeared in Paris.

Onimeter (survey instrument) 1869. *Inv.* Eckhold (Ger) as combination of theodolite and level.

Opera. Created, as an attempt to reconstruct the formula of Greek meliopeia, by Peri and Monteverde (1567–1643). 1656. *Intro.* in England ("Siege of Rhodes.") 1669. *Intro.* in Paris by Abbot Perien. 1710. First Italian opera in England ("Almahide").

Opthalmoscope 1850 *Inv.* Hermann von Helmholtz (Ger) (1821–94). 1849. Crude device used by Charles Babbage.

Oratorio *c.* 1550. Musical style originated by Philip Neri. 1732. First performed in England at Lincoln's Inn Fields theatre.

Orchestrion (automatic orchestra) 1789. *Inv.* Abbé Vogler (Fr). 1810. "Organo-lyricon" *inv.* Saint-Paul, of Paris. 1817. "Apollonicon" *const.* by Flight and Robson, London. 1851. Kauffmann *inv.* five self-acting orchestral machines. 1861. Welte, of Vohrenbach *inv.* the "Orchestrion."

Organ *c.* 300 B.C. Ctesibos *inv.* water-blown organ, or hydraulos. 220 B.C. *Inv.* ascribed to an Alexandrian barber. 1791. Organ without bellows *inv.* by Benedictine monk Luzuel. 1851. Henry Willis *inv.* pneumatic control for organ at Crystal Palace. 1856. American vacuum organ (harmonium) *inv.* by Estey, of Attenborough, Vermont, U.S. 1866. Electro-pneumatic organ in use at Salon, Provence, Fr., with magnetically controlled wind-

valves. 1934. Laurens Hammond (U.S.) *inv.* "Hammond" organ. 1939. Laurens Hammond *inv.* "Novachord" organ.

Orlon 1948. *Intro.* by du Pont de Nemours (U.S.).

Orrery 1696. *Inv.* by and named after Lord Orrery.

Orthicon *See* Cathode-ray tube.

Oscillograph 1893. Mirror-type *inv.* by Blondel (Fr). 1897. Mirror-type *imp.* by Duddell.

Osmium (element) 1804. *Disc.* by Charles Tennant (1768–1838).

Osmosis 1748. *Disc.* as taking place through animal membrane into sugar solution by Abbé Nollet (Fr) (1700–70). 1815. Re-*disc.* by Parrot, and also by W. Fischer in 1822. 1827. R. J. H. Dutrochet measured osmotic pressure and experimented with endosmosis and exosmosis. 1848. Basic principles explained by Baron Liebig. 1867. Membranes for osmosis prepared by Traube. 1877. Rigid membranes prepared by W. Pfeffer. 1898. Rigid-member osmometers perfected by Naccari.

Osteopathy Named and first practised by Andrew Taylor Still (U.S.) (*b.* 1828), of Virginia. 1902–3. Manipulative system brought to Britain by J. Dunham, L. Willard Walker, and Franz Joseph Horn.

Otaphone (deaf-aid) 1836. Webster *inv.* an *imp.* otaphone.

Otosclerosis First recognized by Joseph Toynbee, of Lincoln (1815–66).

OTTO, Dr. Nikolaus A. (Ger) (1832–91) Conducted early experiments with gas-engine, thus helping in the transition between it and the petrol engines of Benz and Daimler.

Overlaying (metal) 1600 B.C. Overlaying in gold, silver and copper practised in Ancient Egypt.

Ovum (anatomy) 1827. Mammalian ovum *disc.* by Ernst von Baer, of Koenigsberg (1792–1876).

Oxalic Acid 1862. First obtained by Dr. Dale.

Oxygen 1727. Accidentally made by Stephen Hales (1677–1761). 1771. *Disc.* by C. W. Scheele and named (1774) by Antoine-Laurent Lavoisier (Fr). 1799. First administered by Dr. Thomas Beddoes and Sir Humphry Davy at the Pneumatic Institute, Bristol. 1877. First liquefied by Louis Cailetet (Fr). 1887. First high-pressure oxygen cylinders *inv.* 1889. First commercially liquefied by Karl von Linde (Ger) (1842–1934). (*See also* Refrigeration; Gases.)

Ozone 1785. Van Marum first noticed change in oxygen made by passage of electric spark. 1840. Dr. Christian Frederick Schönbein, of Basle *disc.* ozone. Ozone *disc.* to be magnetic by Alexander E. Becquerel (Fr) (1820–91). 1904. Water first ozonated at Nice, France.

P

PACINOTTI, Antonio (It) (1841–1912) 1860. *Inv.* first ring-wound dynamo armature.

Packing (gland and piston) 1674. Sir Samuel Morland (1625–95) *inv.* hat-leather packing rings for his metal plunger-pump. 1797. Rev. Edmund Cartwright (1743–1823) *inv.* first expanding metallic piston. 1798. Cupped leather packing *inv.* by Henry Maudslay (1711–1832) for Joseph Brama's hydraulic press. 1813. Bryan Donkin, of Penzance *inv.* double-expanding piston-ring used by Maudslay and Perkin when modified by them. 1820. William Jessop, of Butterley Ironworks *inv.* spiral-spring packing ring for pistons. *c.* 1825. John Barton *inv.* metallic piston-packing. 1852. Spit piston-ring *inv.* John Ramsbottom. 1856. David Joy *inv.* split-spiral piston-ring. 1880. Metallic packing *intro.* Britain by U.S. Metallic Packing Co., of Philadelphia. (*See also* Piston rings.)

Paddle 1752. Paddle-wheel (of modern type) *inv.* by Bernouilli (It). 1785. Chain-paddle *inv.* by Fitch (U.S.). (*See also* Waterwheel and mill.)

PAIN, Albert C. (1856–92) 1890. *Inv.* pressurized engine oil lubrication system.

Painting, Water-colour *Intro.* end of 18th cent.

Palladium (element) 1803. *Disc.* Dr. W. H. Wollaston (1766–1828).

Pallas (planet) 1802. *Disc.* Heinrich Wilhelm Matthäus Olbers (Ger) (1758–1840).

Panemore (globular windmill) 1655. Mentioned by the Marquis of Worcester in his *Century of Inventions*, No. 15. Later re-*inv.* by Desquinemare (Fr).

Panning (metal recovery) 3000 B.C. Practised by goldminers of Ancient Egypt.

Panorama 1788. *Inv.* by Robert Barker, of Edinburgh.

Pantograph 1603. *Inv.* by Christopher Scheiner (Ger). 1821. *Imp.* by Prof. Wallace and called "Eidograph."

Pantomime 364 B.C. *Intro.* into Ancient Rome. 1530. *Intro.* as "modern" *inv.* into Italy by Ruzzante.

Paper 2000 B.C. Papyrus used in Egypt. 190 B.C. Parchment superseded papyrus. A.D. 105. *Inv.* (?) by Tsai-Hun (Ch), of Leiyang. 600. Made from cotton in China. 793. First paper-mill erected at Baghdad. 1290. Water-power applied to pulping rags at Ravensburg, Yorkshire. 1300. Paper first made from rags. 1690. White letter-paper produced. 1719. A. F. de Reaumur (Fr) proposed that paper could be made from wood, but never tried out the idea. *c.* 1732. Clergyman Jacob Christian Schaffer (Ger), also suggested wood for papermaking. 1855. Esparto-grass for paper-making began to be imported. 1857. Vegetable parchment paper *inv.* 1873. Chemically produced woodpulp developed. 1895. Yarn first *prod.* from paper (xylolin) by Emil Claviez, of Saxony. 1903. Kraft paper (sulphated) *prod.* in Norway.

Paper-making Machines 1798. Continuous machine using no manual labour *inv.* François Nicholas Robert, of Paris. (Roberts's *pat.* sold to Didot Saint Leger in 1800, who later sold British *pat.* to Henry and Sealy Fourdrinier. The Fourdriniers and Bryan Donkin (China-clay) *prod.* in 1903 a continous paper-making machine in England.)

Paper-money 1024. *Intro.* in Tabriz.

Papier Mâché 1740. Martin (Ger) learnt technique from Lefevre (Fr). 1745. First made in England by John Baskerville (1706–75).

PAPIN, Denis (Fr) (1647–1712) *inv.* pressure-cooker (digester) and the safety-valve, among other *invs.*

PARACELSUS (Phillipus Aureolus Theoprastus von Hohenheim) (Swit) (1493–1541) Gave first stimulus to the careful classification and examination of all natural substances and founded the Paracelsian School.

Parachute *c.* 1495. Idea sketched by Leonardo da Vinci (1452–1519). 1500. Used in Siam (?). 1617. First *illus.* of parachute appeared in Verantius's *Machinae Novae*. 1783. First tried by Normand (Fr), who jumped from a house

window with a 30 in. diameter umbrella. 1897. Garnerin (Fr) made a parachute descent at Paris with a 23 ft. diameter parachute.

Paraffin 1830. First obtained and named by Georg von Richtenbach (Ger) (1788–1869). 1847. Procured from shale oil by James Young. *c.* 1856. First *prod.* by distillation by Dr. Gesner, of London.

Parallax of Moon 1760. Determined by J. J. Laland (Fr) to be 57° 15′.

Parallax of Sun 1822. Determined by J. F. Encke (Ger) to be 8.57″.

Parity, Conservation of 1927. Law enunciated by E. Wigner.

Parking-meter 1935 (July 16). First one installed in Oklahoma City, U.S. by the Parking Meter Co. *Des.* by Carl C. Magee and known as the "Park-O-Meter."

PARSONS, Charles Algernon (1854–1931) 1884. *Des.* and *const.* the first practical steam turbine. 1894. Built first turbine-propelled ship. (*See* Turbine, steam.)

Parthogenesis (biology) First *disc.* by Antonie van Leeuwenhock (Hol) (1632–1723), in aphis. 1745. Re-*disc.* Charles Bonnet, of Geneva (1720–93).

PASCAL, Blaise (Fr) (1623–62) *Inv.* the wheelbarrow (*q.v.*). 1639. First enunciated theorem of conic sections later known as "Pascal's Theorem." 1641. *Inv.* calculating machine (*q.v.*). and an omnibus.

PASTEUR, Louis (Fr) (1822–95) Biological chemist and bacteriologist. *Disc.* the part germs play in causing disease. 1848. Demonstrated that polarization in crystals (optical activity) was a molecular property. *Disc.* bacillus of rabies.

Paul's Tube (surgery) 1892. *Inv.* Dr. Frank Thomas Paul (1851–1941).

Paving, Road 312 B.C. Paved Appian Way in Rome *const.* 9th cent. Streets of Cordova, Spain, paved. 1185. Rough stone slab paving used in Paris to reduce dust. 1533. London first paved. 1717. Le Large (Fr) *inv.* paving machine. 1824. Joseph Aspdin made Portland cement (*q.v.*). 1827. Concrete road *pat.* by Hobson. 1834. Rolled granite chips first grouted with tar-mac by John Loudon Macadam (1756–1836). 1839. Wood-block paving first tried. 1870. Asphaltic matter first used for roads by de Smelt (Fr) and asphalt macadam by Clifford Richardson in 1894. (*See also* Roads.)

Pearls, Artificial 1716. Formula for making *inv.* by Jannin, of Paris.

PECQUER, Onisiphore (Fr) 1827. *Inv.* the differential gear.

Pedometer *c.* 1756. *Inv.* by Hohlfeld (Ger). 1799. *Pat.* by Ralph Gouts. 1831. Pocket instrument *pat.* by William Payne, of London.

PELIGOT, Eugène Melchior (Fr) (*b.* 1812) *c.* 1840. With J. B. A. Dumas first recognized the radical "methyl" as being present in methane and wood alcohol.

Pellagra Dietary origin of complaint *disc.* by Goldberger.

PELTIER, Jean Charles Athanase (Fr) (1788–1842)

Parisian watchmaker who, in 1834, *disc.* that a weak electric current could produce a cooling effect when flowing through a thermo-couple—the "Peltier Effect."

PELTON, L. A. (1829–1908) 1880. Devised the impulse water-wheel now bearing his name.

Pen 635. Quills first used for pens according to St. Isidore of Seville. Mid-7th cent. Quill pens in general use. 1748. Earliest record of steel-nibbed pen in written claim of Johann Janssen of Aix-la-Chapelle to have *inv.* it. (Not as popular as quills of grey goose.) 1780. Cylindrical steel pens cut like quills made by Harrison of Birmingham. 1803. Joseph Wise *inv.* metal-nibbed "perpetual pen." 1808. Bryan Donkin *inv.* first two-piece steel pen-nib of No. 23 S.W.G. metal. 1820. Joseph Guillot, of Birmingham *inv.* new steel nib. 1823. Hawkins and Mordan *inv.* nibs of horn and tortoiseshell. 1824. Doughty, of London made gold nibs ruby-tipped. 1830. James Perry, of London *inv.* steel nib with side slits. 1830. Donkin *inv.* "bow" or ruling-pen. 1831. Morden and Brockedon *inv.* "inclined" pen-nibs. 1845. Quills for pens finally superseded. *c.* 1870. Vibrating "electric" pen *inv.* T. A. Edison (U.S.).

Pen, Ball-Point 1888. John L. Loud (U.S. *pat.* ball-point pen. 1938. Brothers Ladislao J. and Georg Biro (Hung) re-*inv.* ball-point pen. (Glycerinated ink for ball-point pens *inv.* by Franz Seech (Aus).)

Pen, Fountain 1663. Samuel

Pepys (1632–1703) used a "reservoir pen." 1740. "Pen without end" *inv.* by Jean Félitité Coulon de Thévenot (Fr) (1754–1813). 1809. First two British *pats.* referring to fountain pens. 1819. John Schaffer *inv.* "Penographic" pen with quill-shaped brass nib. 1833. William Baddeley *inv.* plunger-filled fountain pen.

Pencil 1561. Black-lead pencils made by Konrad von Gesner (1516–65). 1795. Nicholas Jacques Conté, of St. Cénari, Normandy *inv.* pencils of pulverized graphite and potter's clay—the Conté pencil. 1823. Hawkins and Mordan *pat.* ever-pointed pencil. 1863. Johann Faber, of Stein, near Nuremberg *pat.* screw-top, ever-pointed pencil-case.

Pendulum 1657. First applied to clocks by Christiaan Huygens (1629–95), and also between 1600–50 by various inventors, including Galileo. *c.* 1700. Fromanteel, first put pendulum to practical use in England. 1851. J. B. L. Foucault (Fr) (1819–68) demonstrated rotation of earth by pendulum at the Pantheon, Paris. "Gridiron," temperature-compensating pendulum *inv.* John Harrison. (*See also* Clocks and Watches.)

Penicillin (antibiotic) 1929. *Disc.* in fungus Penicillium notatum by Prof. Alexander Fleming. (Original *disc.* made in 1928.) *Intro.* into clinical medicine in 1939 by Sir Howard W. Florey and Dr. Ernst Chain. theory of heat; stating that the product of atomic weight and 1941 (Feb. 12). First injection made.

Pepsin 1835. *Disc.* Theodore Schwann (Ger) (1810–82).

Percussion Principle *disc.* Dr. Alexander Forsyth (1769–1843).

Percussion-cap 1807. Chlorate of potash, sulphur, powdered glass and charcoal mixture used as fulminating powder by Rev. Forsyth, of Aberdeen. 1814. Landscape painter Joseph Shaw *inv.* safe, steel-cap for bullet; a pewter percussion-cap in 1815; and a disposable copper one in 1816.

Per-iodic Acid *Disc.* by Magnus.

Periscope 1875. First applied to railway trains by W. A. Robinson, of G.W.R., Canada.

PERKIN, Sir William Henry (1838–1907) 1856. *Disc.* first artificial dyestuff—magenta. Also *prod.* first synthetic perfume—coumarin.

PERKINS, Jacob (1766–1849) 1823. *Inv.* flash steam boiler with a working pressure of 1,500 p.s.i.

Perrotine (fabric-printing machine) 1834. *Inv.* by Claude Perrot (Fr).

Perspective First studied by Ucello (It) (*d.* 1432). 1822. First (*sic*) proposed by Jean Victor Poncelet (Fr) (1788–1867). 1832. Theory furthered by Jakob Steiner (1796–1863); whose proofs were revised and *imp.* by Luigi Cremona (It) (1830–1903). 1847. Theory again furthered by von Standt (Ger) (1788–1867).

Perspiration (anatomy) *Disc.* by Santorio (1561–1638).

PETIT, Alexis-Thérese (Fr) (1791–1820) 1819. With P. L. Dulong interlinked the atomic theory of his time with the theory of heat; stating that the product of atomic weight and specific heat of an element was 6.

Petroleum 1540. "Naptha" known to have been transported in Western Europe in flasks on carts. *c.* 1600. Bitumen-shale treated at Pitchford-on-Severn, Shropshire, and a black varnish ("Welsh lacquer") *prod.* from shale at Pontypool. 1863. Petroleum *disc.* in U.S. *See also* Gas, natural. 1922. Tetra-ethyl lead as an anti-knock additive to petroleum *disc.* by Boyd and Thomas Midgley Jnr., of General Motors, U.S.

Petroleum, Catalytic cracking of 1915. Aluminium catalyst *inv.* by A. M. McAfree and C. F. Cross. *c.* 1930. Nickel catalyst and "fixed bed" process, *inv.* Eugène Houdry (Fr). 1941. Fluid process *intro.* by Standard Oil Company, U.S.

Petroleum, Synthetic Friedrich Bergius *dev.* process involving use of powdered coal and heavy oils. 1925. The wholly synthetic Fischer-Tropsch process *dev.* 1927. First commercial plant built in Leuna Werke, Mersburg, Germany. 1934. "Kogasin" produced from coke, air and water by Ruhr-Chemie, Germany.

PEURBACH, John (1423–61) Pioneer mediaeval German astronomer who proved that Ptolomaic astronomy had been fully absorbed in Western thought by his book *Theoricae novae planetarum.*

Phakascope (optics) *Inv.* Hermann von Helmholtz (1821–94).

Phasmatrope (optics) 1870. *Inv.* by Henry Heyl.

Phenacetin (drug) 1887. *Disc.*

Phenol (carbolic acid) 1834. *Disc.* and named by Carl David Tolmé Runge (Ger) (1856–1927). 1846 (other authorities). *Disc.* Augustin Laurent (1807–53).

Phonautograph 1856. *Inv.* Léon Édouard Scottde Martinville (Fr).

Phonograph 1877. *Inv.* Thomas Alva Edison (U.S.) (1847–1931).

Phosgene 1811. *Disc.* Sir Humphry Davy (1778–1829).

Phosphorescence 1603. *Disc.* by Vicenzo Cascariolo of Vienna when calcining "Bolognese Stone." 1675. Electroluminescence *disc.* by Jean Picard (Fr) when shaking mercury in the dark. 1676. Thermoluminescence *disc.* by Johann Sigismund Elsholtz when heating fluorspar. 1790. Crystalloluminescence *disc.* J. G. Pikel and J. Schönwald. 1838. Fluorspar found to be luminescent by Sir David Brewster (1781–1868). 1845. Quinine sulphate *disc.* to be luminescent by Sir John Frederick William Herschel (1792–1872). *See also* Luminescence and fluorescence.

Phosphoretted Hydrogen (PH₃) 1783. *Disc.* by Gengembre.

Phosphoroscope *Inv.* by Alexander Edouard Becquerel (1820–91).

Phosphorus 11th cent. *Disc.* by Alchid Bechir (Arab). 1669. First made accidentally by Brandt, of Hamburg, by heating urine, coke and sand. 1774. *Isol.* by Gahn (Swed) and C. W. Scheele (Swed). (Also reputed *disc.* by Johann Künkel (Ger) (1630–1703).)

Phosphorus Oxychloride *Disc.* by Würtz.

Phosphorus Trifluoride *Disc.* Henri Moissan (1852–1907).

Photo-electric Cell 1867. Hertz experimented with photo-electric effects. 1873. Photo-electric properties of selenium *disc.* by George May, a telegraphist at Valentia, Eire. Results confirmed by Willoughby-Smith and W. G. Adams. 1888. Hertz *const.* first selenium photo-electric cell.

Photography 1727. Effect of light on silver chloride *disc.* by Dr. Johann Heinrich Schultz (Ger). 1826. Joseph Nicéphore Nièpce (Fr) (1765–1833), of Châlon-sur-Marne, took world's first picture there. 1839. L. J. M. Daguerre (Fr) perfected production of silver photographic image on a copper plate. 1841. William Henry Fox-Talbot (1800–77) *inv.* paper-negative "Callotype." 1847. Nièpce made first sensitive plate of silver iodide on albumen. 1850. Collodion wet-plate *inv.* 1851. Scott-Archer used collodion in his wet-plate process. 1864. B. J. Sayer and W. B. Bolton *inv.* silver bromide emulsion in collodion applied to dry-plate. 1871. R. L. Maddox *inv.* use of silver bromide emulsion in gelatine. 1884. Roll-film *inv.* by W. H. Walker. (*See also* Camera.) 1885. George Eastman (U.S.) (1854–1932) *pat.* machine for making continuous photographic film. 1888.

George Eastman placed first roll-film camera on market.

Photography, Astronomical 1850. W. C. Bond (U.S.) (1789–1859) made first successful experiment at Harvard, U.S., by taking photograph of the moon. 1857. Bond first photographed a double star. 1858. Warren de la Rue took first picture of scientific value of the sun. 1882. John William Draper (1811–82) and H. C. Vogel took picture of great nebula of Orion.

Photography, Colour 1890. Lippmann, of Paris, photographs solar spectrum in colour. 1891. Lippmann photographed stained-glass window, fruit and flowers in colour. 1912. Dr. Rudolph Fischer (Ger) *pat.* basic principles of dye-coupler colour process (the "Kodachrome" process). 1923. Leo Godowsky Jnr. and Leopold Mannes (U.S.) *prod.* crude picture containing all colours on a three-layer type of film. 1935. Godowsky and Mannes *imp.* system marketed by Eastman Kodak as "Kodachrome."

Photogravure 1852. *Inv.* Fox-Talbot.

Photometer, Photometry 1080. Sufi (Abderrahman) (Arab) *imp.* methods of photometry of stars. 1760. Science founded by J. H. Lambert (Ger). Photometer *inv.* by Count Rumford (1753–1814). Other types *inv.* by Arago, and Wheatstone. 1844. "Greasespot," or shadow photometer *inv.* R. W. Bunsen. 1889. Detail photometer *inv.* Otto Lummer and Eugen Brodhuhn.

Photons Albert Einstein (1879–1955) suggested that light consisted of bundles of energy—now called photons. (*See also* Investigations of Prince Louis de Broglie and Edwin Schrödinger.)

Photophone 1880. *Inv.* in U.S.A.

Photosynthesis 1779. *Disc.* by engineer Jan Ingenhousz (Hol) (1730–99), while working in London. (*See also* Chlorophyll.)

Photozincography 1800. Perfected by Col. James. (*See also* Printing.)

Phrenology 1785. First proposed as science of craniology by Dr. Gall (Ger) (1758–1828). 1812. Spurzheim (Ger) expanded on Dr. Gall's ideas. 1819. Combe published first English book on phrenology.

Phtalein (dyes) 1871–87. *Disc.* by Adolf von Bayer (Ger).

Physharmonica (*See* Organ, American.)

Pianoforte 1709. *Inv.* Bartolommeo Christofori (It), and first brought to England from Rome by monk, Father Wood. 1716. Marius, of Paris, *inv.* upright pianoforte. *c.* 1830. Pleyel, of Paris, *inv.* iron-framed pianoforte. 1837. Jonas Chickerling (U.S.), of Boston, Mass., *intro.* iron-framed pianoforte in U.S. Sebastian Erard (1752–1831) made many improvements to pianoforte. 1850. Paper-manufacturer Andrew Dimoline, of Bristol, *inv.* compensating pianoforte action.

Piano-player, Player-piano Mechanical piano *inv.* Nicolo Frabris (1739–1801). Muzio Clementini (It) (1752–1832) *inv.* a "self-performing piano." 1848. Höhlfeld (Ger) *inv.*

"music machine recording on paper." 1876. Piano player *inv*. John McTammany (U.S.) (*pat*. 1881). 1896. Edwin S. Votey (U.S.) *inv*. first practical piano player and coined word "Pianola." (*See also* Barrel-organ and "Orchestrion.")

PIAZZI, Guiseppe (It) (1746–1826) 1801. *Disc*. first minor planet Ceres.

PICARD, Jean (Fr) (1620–82) 1675. *Disc*. Electro-lumines-cence.

PICKARD, John (n.d.) *Pat*. the crank in late 18th cent. (English *Pat*. No. 1263.)

Picric Acid 1771. First *disc*. P. Woulf. 1788. J. F. Hausmann re-*disc*. picric acid and used it as a dye. 1871. Sprengel *disc*. picric acid to be explosive. 1886. Picric acid first used as explosive. 1888. Picric acid *intro*. into England and used as "Lyddite." (*See also* Explosives.)

PICTET, Marc-Auguste (Swit) With C. W. Scheele and Edmé Marriotte carried out early experiments in radiant heat.

PIERRE DE MARICOURT (*c*. 1250) Pioneer of systematic scientific experiments in magnetic phenomena. (*See Epistola magnete*, written by him in 1269.)

Piezo-electricity *Disc*. by Pierre and Jacques Curie (Fr).

Piezometer *Inv*. by Hans C. Œrsted (Dan) (1777–1851).

Pile, Electric *See* Battery, electric.

Piles, Interlocking 1588. *Illus*. by Augustin Ramelli (1531–90).

Pile, Screw *Inv*. Alexander Mitchell and first used to erect lighthouse on the Maplin Sands.

Pilot, Automatic *See* Gyro-stabilizer (aircraft).

Pins 1483. Bone pins first used in England. 1540. Brass pins imported into England from France. 1543. Pins first made in England. 1824. Lemuel Well-man Wright *inv*. pin-making machine.

Pin, Safety 1849. *Inv*. and *pat*. Walter Hunt, of New York— sold *pat*. for £300.

Pinchbeck (alloy) *Inv*. Thomas Pinchbeck (*d*. 1783).

Pipe 1st cent. B.C. Lead and bronze pipes used by Romans. 1790. Lead seamless pipe cast by John Wilkinson, and brass and copper pipe drawn by Charles Green. 1797 Extrusion method suggested, but not practised by Joseph Bramah. 1820. Extrusion method *inv*. by Thomas Burr, a Shrewsbury plumber, using a hydraulic ram to produce lead pipe. 1860. R. Mannesmann (Ger), of Remscheid, conceived idea of making seamless steel tubing. 1885. The brothers Mannes-mann publicize their tube-rolling process.

Pipe, Tobacco-smoking 1850. Condensing type *inv*. by William Edward Stait.

Pipe-line 1874. First one laid by Van Sickle (U.S.).

Piston *Inv*. in the West by Theophile Desaguliers (Fr) (1633–1744). 1590. Screw-oper-ated piston *desc*. by Cyprian Lucar in his book *Lucar solace*.

Piston-horn (music) 1815. *Inv*. Heinrich Stolzel (Ger).

Piston-ring *See* Packing, piston.

Pitot Head 1732. *Inv.* Henri Pitot (Fr).

PIXII, Hippolyte (Fr) (*n.d.*) 1832. *Inv.* first electro-magnetic machine (dynamo) and later same year, at suggestion of A. M. Ampere, fitted it with commuting device and *prod.* direct current.

PLANCIUS, Petrus (Hol) (1551–1622) Did much pioneer work in physics, geometry, astronomy and mathematics with Simon Stevin (Stevinius).

PLANCK, Max 1900. *Prop.* the Quantum Theory.

Plane, Inclined Principle of first established by Simon Stevin, of Bruges (1548–1620).

Plane-table (surveying instrument) 1551. First *desc.* by Abel Foullon, of King Henry II's household.

Planetarium 1749. Clockwork planetarium *inv.* and made by French optician Claude Simon Passement. *c.* 1830. Another *inv.* Thomas Harris Barlow (U.S.). 1923. Projection planetarium *inv.* by Prof. Walther Bauersfeld, of Munich.

Planets, Astronomical Laws relating to the 406–366 B.C. Regular motions of planets first *disc.* by Eudoxus. 1618. Law *disc.* now bearing his name by John Kepler (Ger) (1571–1630).

Planimeter 1814. *Desc.* by J. M. Herman. 1824. *Desc.* by Gonella, of Florence. 1827. *Desc.* by Johannes Oppikoffer, of Berlin. 1849. *Desc.* by Wetli, of Vienna. 1851. *Inv.* (*sic*) by Paul Cameron, and by John Sanginer. 1854. Polar planimeter *inv.* Jakob Amsler of Zurich. 1860. Planimeter *inv.*

(*sic*) by John Dillon, of Dublin.

Planing Machine 1794. In use at Horace Miller's works, Preston, Lancs. 1802. J. Bramah *dev.* planing machine *inv.* by his draughtsman Joseph Clement for use at Woolwich Arsenal (worked by a 90 h.p. steam-engine). Sir Marc I. Brunel *inv.* (*sic*) planing machine. 1814. *Inv.* James Fox, of Derby. 1817. Chain-drive planing machine *inv.* Richard Roberts. J. Wilkinson *inv.* first metal planing machine. 1827. Jointly with William Thompson, a cabinet-maker of Fountainbridge, Edinburgh, Malcolm Muir *inv.* planing machine for floor-boards. 1836. John McDowall *inv.* planing machine adapted for grooving and tongueing floor-boards. 1857. James Nasmyth *inv.* gear-driven planing machine.

Planisphere 1731. Clockwork planisphere *inv.* and made by Abbé Reginald Outher, showing the phases of the moon.

PLANTA, Martin (Swit) (1727–72) Early investigator of electro-statics and electrical instrument-maker.

PLANTÉ, Raimond-Louis-Gaston (Fr) (1834–89) 1859. *Inv.* the lead cell electric accumulator.

Plants, Classification of 1570. Matthias de l'Obel made first attempt at plant classification. 1583. Andrea Cesalpino (It) (1519–1603) arranged plants by flowers and fruits. Kaspar Bauhin (Swit) (1560–1624) distinguished between the idea of genus and species. (*See also* Binominal nomenclature of animals and plants.)

Lobel (1538–1616) *disc.* the two groups of plants—monocotyledons and dicotyledons.

Plants, Fertilization of 1789. *Disc.* by Sprengel as being due to insects. Sex-organs of flowers *disc.* Nehemiah Grew (1641–1712).

Plants, Respiration of A.D. 950. *Disc.* by Al Farabi (Arab). Re-*disc.* Stephen Hales (1677–1761).

Plastic Coating 1924. Albert E. P. Girard and Maurice Roumanzielles jointly *pat.* process for coating yarns with plastic.

Plastic Surgery 1553. Gaspar Taliacotius (It) made artificial noses. 1800. Plastic surgery *intro.* into Europe as "Rhinoplasty" by French surgeon Lucas, who had seen an Indian physician restore a man's nose with skin from the forehead. 1827. The art of plastic surgery revived in France.

Plate, Sheffield 1742. Process *inv.* Thomas Bolsover, of Sheffield, for making buttons. Matthew Boulton (1728–1809) devised method of soldering silver wire to mask edge of plated articles.

Platinum 1735. *Disc.* by the Jesuits in Columbia. 1750. *Isol.* by Watson. 1851. First used for chemical apparatus.

PLATO (429–347 B.C.) Athenian philosopher who first realized the importance of mathematics to science.

Plimsoll Line 1876. *Inv.* and *intro.* by Samuel Plimsoll (1824–98).

PLINY the Elder (A.D. 23–79) Roman natural scientist. Wrote *Naturalis historia.*

Plough Pre-3000 B.C. Used in Egypt and Mesopotamia. *c.* 2000 B.C. Earliest iron ploughshares found in Palestine. (No ploughshares are traceable to Hellenic sources, but socketed and spearheaded types were commonly used in Ancient Rome.) 500 B.C. Curved wooden mould-board plough used in China. 11th cent. Plough *intro.* into Europe. 1523. Wheeled plough noted by Fitzherbert. 1619. David Ramsey and Thomas Wildgoose *pat.* ploughing machine (English *Pat.* No. 6). 1627. William Brouneker, John Aprice and William Parham *inv.* machine for ploughing and tilling worked entirely by two men. 1634. William Parham, John Prewett, Ambrose Prewett and Thomas Dorney *inv.* "engine" for ploughing without oxen. 1707. Iron ploughbreast in use in Essex. 1730. "Rotherham" plough in use. Disney, Stanyforth and Joseph Foljambe *inv.* plough. 1763. Cast-iron plough mould-board *inv.* and made by Small at his Blackadder Works. 1766. Cuthbert *inv.* drawing (traction) plough. 1785. Robert Ransome *inv.* and *prod.* cast-iron, tempered ploughshares. 1800. Richard Lambert *inv.* mole plough. 1800. Plenty *inv.* friction-wheel plough. 1808. Robert Ransome *inv.* plough with replaceable parts. 1837. John Upton took out first steam-plough *pat.* 1853. John Oliver *inv.* chilled iron ploughshares. 1855. "Guideway" ploughing system *pat.* by Halkett. 1867. Phillip Smith *des.* ploughing system for use with

Howard's transverse ploughing traction-engine. *c.* 1890. Disc plough *inv.* (*See also* Traction engine.)

Plumb-line 1100 B.C. Used in Ancient Egypt.

Pluto (planet) 1930. *Disc.* by Clyde Jonbaugh.

Plutonium (element) 1942. *Disc.* by Glen T. Seaborg, E. M. McMillan, J. W. Kennedy, and A. C. Wahl (U.S.) by transformation of uranium 238 by neutron capture. 1948. Problem of making a suitable container for plutonium at high temperature solved. 1949. Plutonium being *prod.* in England. 1950 (July). Atomic plant at Windscale, Cumberland began *prod.* plutonium on a large scale.

Plywood 2800 B.C. *Inv.* in Ancient Egypt, where wood was very scarce; six plies being fastened together by wood pegs. *c.* 1918. Popularized in England through work done by French prisoners of war.

Pneumonia Bacillus *disc.* by Albert Fraenkel (1848–1916), of Berlin.

POINCAIRE, Jules Henri (Fr) (1854–1912) *c.* 1903. Savant responsible for many critical observations on the mathematics of mechanics.

POINSOT, Louis (Fr) (1777–1859) 1803. Savant responsible for the development of the theory of couples of forces, in his work *Elements de statique.*

Points, Railway (*See* Signal interlocking, railway.)

POISSON, Siméon-Denis (Fr) (1781–1840) Savant whose name has remained connected with the law governing the adiabatic changes in the condi-

tion of a gas. (Boyle-Marrotte law relates to the corresponding isothermal conditions.)

Poisson's Brackets (mathematics) 1925. Significance of *disc.* A. M. Dirac (Fr).

Polarimeter 1840. *Inv.* by Jean Baptiste Biot (Fr) (1774–1862). 1842. *Intro.* into the sugar industry by Ventzke (Ger).

Polarized Light 1690. First recognized and explained by Christiaan Huygens (Dan) (1629–95). 1808. E. L. Malus (Fr) (1775–1812) *disc.* practical application of polarized light. 1828. Polarizing prism *inv.* by Scotsman William Nicol. 1845. Michael Faraday *disc.* magnetic rotation of polarized light. 1850. Dr. W. B. Herepath *disc.* polarizing properties of combined iodine and quinine sulphates. 1925. Edwin H. Land (U.S.) *inv.* first practical synthetic light-polarizing material and demonstrated it in 1932. (The chromatic polarization of light *disc.* by D. F. Arago.)

Polarography 1922. *Inv.* Jaroslav Herovský, of Prague.

POLHEM, Christopher (Swed) (1661–1751) *Inv.* and *dev.* the section-rolling, water-driven iron mill, thus transforming the iron-rolling industry.

Politzer's Bag (surgery) 1863. *Inv.* Dr. Adam Politzer (1835–1920), of Vienna.

Polka 1835. *Inv.* in Prague. 1844. *Intro.* into England.

Polonium (element) 1898. *Disc.*

Polyphon (*See* Musical-box.)

Polythene 1933. *Disc.* by P. H. Fawcett. 1936. Pilot continuous process in use. 1938. Full-scale pilot plant in operation. 1939. One mile of submarine cable insulated with polythene.

PONCELET, Jean-Victor (Fr) (1788–1867) Savant who *intro.* use of kilogramme/metre unit in mechanics.

POND, John (1767–1836) One-time Astronomer Royal of England. In 1833 *pub.* catalogue of 1,112 stars—a pioneer work.

PONS, Jean-Louis (Fr) (1761–1831) Astronomer who, in 1818 *disc.* Encke's comet, named after the mathematician who made the *disc.* possible by determining its orbit.

POPHAM, Admiral Sir Home 1803. With Marryat *inv.* the original International Flag Signal Code.

Porcelain 1531. *Intro.* into England. 1687. Von Tschishaus made porcelain from kaolin. 1705. Johann Freiderich Böttger (1685–1719) usually credited with *disc.* of making porcelain.

PORTA, Giovanni Battista Della (It) (1535–1615) Scientist and pharmaceuticist sometimes credited with formulating the idea of the telescope, which was said to have been made in Italy in 1590 before finding its way to Holland.

Portland Stone 1616. First quarried on the Isle of Portland, Dorsetshire.

Positron (absolute particle) 1932. *Disc.* by Carl D. Anderson (U.S.) (*b.* 1905) and Prof. P. M. S. Blackett (U.K.), independently.

Post-card 1862 (Nov. 1). First issued in U.K. 1882 (Nov. 2). First reply-paid post-card in U.K.

Potassium (element) 1807. *Disc.* by Sir Humphry Davy (1778–1829).

Potter's Wheel 3250 B.C. Evidence of vases made on potter's wheel in Sumeria. 3000 B.C. In use in Palestine. 2750 B.C. First appeared in Ancient Egypt (III Dynasty). 2000 B.C. Appeared in Crete. 1800 B.C. In Greece. 750 B.C. In Italy. 400 B.C. Appeared in Rhine and Upper Danube river-valleys. 50 B.C. First appeared in Southern England. A.D. 1550. First evidence of use of potter's wheel in the Americas.

POUILLET, Claude-Servain (Fr) (1791–1868) 1837. With Sir John Herschel was first to measure strength of sun's radiation.

POULSEN, Valdemar (Den) *Inv.* undamped arc radio transmitter and the first tape sound recorder.

Powder, Smokeless 1881. First *prod.*

POWER, Henry (1623–68) 1649. *Disc.* capillary channels between the arteries and veins of the human body. (*See also* Malpighi and Harvey.)

Power Station, Electric 1865. Hydro-electric generating system *prop.* by Cazel (Fr). 1866. Felice Marco (It) *pat.* principle of electric generation by water-power. 1881 (Oct. 12). First public hydro-electric plant brought into use at Godalming, Surrey. 1881. T. A. Edison *des.* and *const.* first steam-driven power-station—New York. 1882. Edison *des.* and *const.* first steam-driven power-station in England—Holborn, London. 1882. First hydro-electric station in U.S. (for D.C. output)—Appleton, Wis. 1883. Giant's Causeway Tramway hydro-electric station opened at Walkmills, Co. Antrim, Ireland. 1889. First alternating currant hydro-electric plant erected at Oregon City, U.S.

Power Transmission (automobile) 1901. Benson *inv.* fluid-coupling drive. 1904. Centrifugally operated, two-speed automatic power transmission fitted to U.S. Sturtevant car. 1906. Dr. H. Föttinger *inv.* fluid flywheel drive. 1924. Alan Coates (Scot) *inv.* convertor coupling. 1930s. Ferguson *inv.* convertor coupling transmission. 1948. Buick "Dynaflow" epicyclic-convertor coupling *intro.* (Types of automatic transmission *inv.* by H. F. Hobbs (mechanical), and by Jacob Rainbow (U.S.) (Smith's electro-mechanical system).) (*See also* Fluid flywheel, Compressed air transmission, Belt transmission, etc.)

Power Transmission (belt and rope) 1750. Richard Arkwright (1732–92) used rope power transmission on cylindrical pulleys, with friction-wheels and toothed-wheels on his spinning frame. James Hargreaves (*d.* 1778) used rope power transmission for steadily driving his spinning jenny. 1800–08. Marc Isambard Brunel used convex pulleys and belt-drive, together with metal-to-metal cone clutches in machines he *des.* for Portsmouth Dockyard. 1850. M. Hirn (Fr), of Colmar, Alsace, *des.* water-wheels with pulleys and endless wires to transmit and use power of waterfalls at a distance. Scheme shown at Paris in 1862. 1856. Rope transmission *intro.* (*sic*) by James Combe, of Belfast. Leonardo da Vinci credited with *inv.* flat belt, crossed belt, and linked chain drives.) (*See also* Chain-drive and gears.)

Power Transmission (compressed air) 1799. George Medhurst (1759–1829) transmitted water-wheel compressed air at 210 p.s.i. to a compressed air motor in a mine. 1803. William Murdock (1754–1839) used compressed air power transmission for ringing his door-bell. 1839. Compressed air power transmission used by London and Birmingham Railway engineer to ring signal-bell between Euston and Camden Town Stations, London. 1861. Compressed air power transmission used to drive rock-drills by Sommeiller (Fr) in driving Mont Cenis railway tunnel.

Power Transmission (electric) 1733. Electricity (static) transmitted 1,256 ft. along a wet thread. 1820. A. M. Ampere (Fr) (1775–1836) suggested electric needle telegraph (*q.v.*). 1870. Froment, of Paris, drove solenoid-operated beam-engine at a distance. 1873. Z. T. Gramme first demonstrated electric power transmission over $\frac{1}{4}$ mile at Vienna Exhibition. 1882. Marcel Duprez built first long distance electric power transmission from Munich to Miesbach (37 miles). Electric power transmission first used in mines. 1884. Duprez builds 25 mile direct current line from

Criel to Paris. 1886. First alternating-current transmission line erected from Cerci, *via* Tivoli, to Rome.

Praesodymium (element) 1885. *Disc.*

Precipitation, Gas 1884. Sir Oliver Lodge re-*disc.* phenomenon that electric discharge through smoke precipitates the smoke. 1904. Dr. F. G. Cottrell (U.S.) repeats Lodge's experiments.

PREGL, Fritz (Ger) (1869–1930) 1910–16. *Dev.* technique of microanalysis.

Preservative, Timber 1795. Vacuum process of impregnation *inv.* by Sir Samuel Bentham. 1828. "Kyanization" process *inv.* by John Howard Kyan (*pat.* 1832).

Press, Power 2nd–1st cent. B.C. Screw-press used for wine and olive oil in Ancient Greece. 1553. Antoine Brucher (Fr) *inv.* coining press. 1664. Blaise Pascal (Fr) (1623–62) proposed hydraulic press. 1790. Fly-press *inv.* by Matthew Boulton (1728–1809). 1796. Hydraulic press *inv.* Joseph Bramah (1748–1814). 1799. Copying-press *inv.* James Watt (1736–1819).

Press-stud (fastener) 1860. *Inv.* by John Newnham, of Birmingham.

Pressure-cooker (autoclave) *Inv.* by Denis Papin (Fr) (1647–1712).

PREVOST, Pierre (Fr) (1751–1839) Astronomer who, with William Herschel, in 1783, determined the so-called solar apex—the point in the heavens towards which the solar system is moving through space.

PRIESTLY, Joseph (1733–1804) English chemist. *Disc.* carbon monoxide, nitrous oxide, nitric oxide, nitrogen, and oxygen gases; the latter on Aug. 1, 1774.

Printing Press, Hand Printing pioneered in Germany by Johannes Gensfleisch Gutenberg (*c.* 1397-1468) 1423. Laurens Janszoon Coster, of Haarlem, printed books using thick ink. 1440. J. G. Gutenberg, of Mainz, printed Vulgate Bible. 1452. Peter Schöffer cast first metallic type in matrices and used it on 2nd edition of the Vulgate Bible. (*See* Type.) (By 1500 there were 1,050 printing-presses in Europe: in Italy, 532; Germany, 214; France, 147; Spain, 71; Holland, 40; other nations, 46; of which seven were in London, three in Westminster, two in Oxford, and one in St. Albans.) 1800. Earl Stanhope *inv.* the "Stanhope" hand-press. 1817. George Clymer, of Philadelphia, U.S. *inv.* the "Columbian" press. 'Albion" press *inv.* R. W. Cope, of London. (A portable hand printing press was *inv.* by Abbé Rochon (Fr).)

Printing Machine, Power 1804. Saxony printer Kœnig *inv.* power printing machine. 1814 (Nov. 28). First machine-printed copy of *Times* appeared. 1818. Augustus Applegath and Cowper *pat.* cylinder machine. 1830. First platen machine *const.* 1837. Thomas Edmondson (1792–1851) *inv.* railway ticket printing machine. 1848. Hoe rotary machine *intro.* into Europe (Paris) from U.S., and into England nine years later.

1851. Thomas Nelson, of Edinburgh *inv.* continuous rotary machine. 1860. Automatic feed *inv.* by C. Sprye, of St. Leonards-on-Sea, Sussex. 1868. Walter *inv.* rotary machine.

Printing, Colour 1455. John Fust (Faust) attempts colour printing. 1820. Chromo-xylography (wood-block) in use. Later *dev.* by William Savage and George Bolton. 1849. Colour printing *dev.* by G. C. Leighton. 1889. Colour printing explained to London printers. (*See also* various colour printing methods: Lithography, Photo-lithography, etc.)

Printing, Copper-plate 1450. Idea conceived in Germany and supposedly *inv.* by T. Fimguerra (1410–75?). (*See also* Blocks, Printing, Zincography, etc.)

Printing, Textile (machines) 1676. Frenchman sets up textile printing works at Richmond, Surrey. 1756. *Inv.* by Bonvallet, of Amiens. 1764. Textile printing *intro.* into Lancashire. 1780. *Inv.* Roland (Fr). 1783. Thomas Bell used engraved metal cylinders for cotton printing. 1785. Bell's roller printing process *imp.* by Nicholson. 1791. *Inv.* Rombillard, of Paris. *Inv.* by Oberkampf and Samuel Widmer, a Swiss chemist. (*See* Perrotine.)

Probability, Theory of First studied by Cardanus, then by Pierre de Fermat (Fr) (1605–65). 1814. Further *dev.* by Laplace and Pierre Simon (Fr) (1749–1827).

Probe (surgery) Porcelain-tipped probe *inv.* by Auguste Nélation (Fr) (1807–73) and first used on Garibaldi to trace bullet by lead mark.

Promethium (element) 1945. *Disc.* jointly by J. A. Marinski, L. E. Glenedin (U.S.), and C. D. Corydell.

Prontosil (drug) 1910. *Disc.* Gerard Domagk (Ger). 1932 synthesized by Klöver and Mietzsch (Ger).

PRONY, G. F. C. M. R. de (1755–1839) *Dev.* brake-type dynamometer.

Propeller (helix) *c.* 400 B.C. *Inv.* ascribed to Archytas of Tarentum. 180 B.C. Hot-air-driven propeller lamp-cover in use in China. A.D. 1505. First applied to air in the West by Leonardo da Vinci (1452–1519); but his note-books not *pub.* until 1797. 17th cent. Hot-air-driven screw appears in Western Europe. 1752. Daniel Bernoulli proposed propeller. 1768. Propeller for ship propulsion *prop.* by engineer Alexis Jean Pierre Paucton (Fr) (1736?–98). 1770. James Watt (1736–1819) sketched his "spiral oar" in a letter to a friend. 1785. *Inv.* (*sic*) by Joseph Bramah, and *pat.* 1794. Lyttelton *pat.* propeller. 1796. John Fitch (U.S.) experimented with propeller at New York. 1802. Capstan-worked propeller fitted to transport ship *Doncaster* by Edward Shorter. 1804. John Cox Stevens (U.S.) drove a steamboat at 7 knots an hour by two narrow-bladed propellers at stern. 1815. Richard Trevithick *pat.* propeller. 1826. Joseph Ressel (*d.* 1857), of Bohemia, used a two-man-power propeller-driven boat on the Danube. 1829. Ressel fitted a 2 ft. diameter propeller to a Trieste boat with a 6 h.p. engine and steamed out to sea. (All further trials of boat stopped by police.) 1836. Propeller independently *inv.* (*sic*) by Francis Pettit-Smith and John Ericsson (Swed). 1838. S.S. *Archimedes* the first successful propeller-driven steamer (237 tons). 1838. T. Lowe *pat.* a propeller and tries it on royal yacht *Fairy*. 1842. Bennet Woodcroft *imp.* propeller by increasing its pitch at blade-tips. 1843. Screw propeller adopted by British Navy. 1844. Propeller-driven ship *Great Britain* built—first propeller ship to cross Atlantic Ocean. 1848. William Maudslay *inv.* variable pitch ship propeller fitted to S.S. *Bosphorus* and used on England–Cape Town route. 1849. Robert Griffiths *imp.* ship propeller by enlarging the boss and reducing the blade-tip area.

Propulsion, Jet 1420. Giovanni da Fontana sketched jet-propelled fish, birds and a rabbit. Also proposed its use for measuring water depths and air heights. 1495. Francesco di Giorgio *des.* jet propulsion petards on wheels and floats. (*See also* Aircraft, and Ship.)

Protein 1838. Term first used ("Protéene") by Gerard Johann Mulder (Hol) (1802–80). 1882. Emil Fischer (Ger) (1852–1919) pioneered modern protein chemistry.

Protoactinium (element) 1917. *Disc.*

Protoplasm 1853. Protoplasm (primitive form of living matter) *disc.* Hugo von Möhl.

PROUST, Joseph Louis (Fr) (1754–1826) Early chemist who disagreed with C. L. Berthollet's theory of chemical compounds as *prop.* in *Recherches sur les lois de l'affinité* (1798).

PROUT, William London doctor who started a hypothesis on the unity of matter in 1815.

Protractor 1801. Three-armed protractor *inv.* Capt. Joseph Huddart.

Prussian Blue 1704. *Disc.* by German dyer Diesbach by accident.

Prussic Acid 1782. *Disc.* by C. W. Scheele (Swed) (1742–86).

Psycho-analysis *See* Subconscious mind.

P.T.F.E. (polytetrafluoroethylene) Used for making frictionless (unlubricated) bearings.

PTOLEMY, Claudius (A.D. 2nd cent.) In his *Almagest* dealt with trigonometric and goniometric principles, together with tables of chords. Pioneer Greek astronomer.

Ptolemaic System (astronomy) A.D. 140. System perfected by Claudius Ptolemy.

Ptyalin 1841. Obtained (from saliva) by Louis Mialhe (1807–1886).

Pulley *c.* 700 B.C. Probably used in connection with chain-of-pots pump to irrigate Hanging Gardens of Babylon.

PULLMAN, George M. (U.S.) (1831–97) 1858. Perfected the railway sleeping-car of 1839 and *des.* the luxury railway coach bearing his name.

Pulsometer *See* Pump.

Pump *c.* 1485 B.C. Supposedly *inv.* by Grecian Danaus. 700 B.C. Chain-of-pots pump on rope used to water Hanging Gardens of Babylon. 3rd cent. B.C. Ctesibus *inv.* plunger pump of which specimens dating as late as 1st cent. B.C. have been found. 2nd cent. B.C. Chain-of-pots for baling mentioned in Egyptian papyri. *c.* 100 B.C. River current-driven wheel-of-pots applied in Egypt during the Roman period. A.D. 1st cent. Square-pallet chain-and-rag pump used in China. 1456. Italian artist Pisanello depicts piston pump powered by a water-wheel and operated by two simple cranks and connecting-rods. 1556. Georgius Agricola (1494–1555) *desc.* in *De Re Metallica* multi-stage reciprocating bucket pumps driven by water-wheels; horse-operated, gear-driven rag-and-chain pumps in series to drain mines up to 600 ft. deep; also mine ventilating pumps, centrifugal fans and water-driven bellows for furnace blowing. 1615. S. de Caus (1576–1630) steam pump in which steam drove out the water. 1616. G. Finugio *illus.* self-acting bucket pump. 1618. R. Fludd (1574–1637) *desc.* automatic free-piston pump to raise clean water by power of dirty water. 1634. Hannibal and Vyvyan, of Cornwall, *inv.* (English *Pat.* No. 67) engine for draining mines. 1635. John Bate *pub.* description of leather plunger pumps and valves. (*See also* Packing, piston.) 1636. D. Schwenter *illus.* two types of gear-wheel pumps. 1654. G. von Guericke made pump and cylinder in which vacuum caused piston to do work. 1672.

Square-pallet chain pump *intro.* into Europe. 17th cent. Air compression vessels for pumps *intro.* in Germany. 1674. Sir John Morland (1625–95) *pat.* plunger pump with metal plunger and hat-leather packing. 1698. Capt. Thomas Savery (1650–1715) *prod.* steam pump of "Pulsometer" type. 1712. Capt. T. Savery installed one of his pumps at Camden House, Kensington, London. (*See also* Newcomen's steam pumping-engine.) 1716. P. de la Hire (Fr) (1640–1718) *desc.* double-acting piston pump. 1724. P. Leupold (1674–1777) *desc.* two-cylinder high-pressure steam pump. 1735. Water-pressure pump used in France. (*See also* Ram, hydraulic.) 1750. Spiral air-lift pump *inv.* by tinworker Andreas Wirz, of Zurich. 1765. W. Westgarth *intro.* water-pressure pump at Coal Cleugh, Northumberland. *c.* 1767. Smeaton *des.* bucket mine pump. 1796. Boswell *desc.* first pump of modern *des.* on principle of Heron's fountain, in Hachette's *Traité élémentaire des machines.* 1840. Henry Howarth (U.S.) *inv.* double-acting, reciprocating steam pump. 1841. Direct-acting steam pump. *inv.* 1857. Double-acting steam pump *inv.* 1872. "Pulsometer" type steam pump *inv.* (*sic*) by H. Hall, of New York.

Pump, Centrifugal 1732. First record of centrifugal pump by Le Demour (Fr). 1783. Thomas Erskine *inv.* centrifugal pump (*pat.*). 1818. "Barker's Mill" (*q.v.*) type of pump *prod.* in Massachussets, U.S. 1831. Disc-type centrifugal pump first *prod.* by Blake, of Connecticut, U.S. 1839. Massachusets centrifugal pump *inv.* (*sic*) by Andrews, of New York. 1846. Valve-disc-type centrifugal pump *pat.* bought by John Gwynne and *intro.* into England. 1846. Andrews's further patents bought by Gwynne and *pat.* in England in 1851. 1848. Appold began making Lloyd-type centrifugal pumps of fan type. 1854. Gwynne *pat.* vertical-shaft centrifugal pump. 1851. Clune *inv.* rotary centrifugal pump. 1876. Prof. Osborne Reynolds *inv.* multistage centrifugal pump without guide-vanes (*pat.* by Gwynne). (Other types of centrifugal pumps were *inv.* 1841, Whitlaw; 1845, Sir Henry Bessemer; 1846. von Schmidt; 1849 and 1850–51, Bessemer *imp.* centrifugal pumps.)

Pump, Cog-type 1636. D. Schwenter *illus.* two types of gear-pump in his *Deliciae physico-mathematicae*. 1823. "Rangeley" pump, with two fluted, cogged rollers *inv.* by Dixon and Rangeley. 1865. Cog-wheel chain pump *inv.* by Jean Bastier (Fr).

Pump, Gas 1906. *Inv.* by H. A. Humphrey (1868–1951) with W. J. Randell. (In this pump, the piston was replaced by the water-column being pumped.)

Pump, Motor-car Petrol 1900. Diaphragm-type petrol pump fitted to Oldsmobile (U.S.) car. 1902. Injector pump fitted by Klaus to his carburettor. 1904. Plunge feed-pump fitted by Gibbon to carburettor. 1905. Petrol feed-pump fitted to

Henroid car. 1911. J. Higginson and H. Arundel *inv.* the "Autovac" petrol feed system. 1928. "A.C." petrol-pump *inv.* by J. N. Morris and M. D. Scott, marketed by A.C.-Spinx Co.

Pump, Vacuum 1650. Otto von Guericke (Ger) (1602–86) *inv.* valveless type. 1659. Robert Boyle and Robert Hooke jointly produce the first English air-vacuum pump. 1865. *Inv.* Hermann Sprengel (Ger). 1913. Rotary vacuum pump *inv.* Dr. Wilhelm Gaede, of Freiburg, Baden.

PUPIN, Michael Idvorsky (1858–1935) Inventor of the acoustic "Loading spool" whereby sound could be transmitted over a distance.

Pycnometer (specific gravity) *c.* A.D. 1000. Applied by Al Biruni (Arab).

Pyocyanase (early antibiotic, originally effective against plague and anthrax, but now abandoned) 1894. *Disc.* by Rudolf Emmerlich, of Munich and Oscar Lowe, of Vienna.

Pyramidon (drug) 1886. Synthesized, for the first time.

Pyrheliometer *Inv.* Claude-Servain Pouillet (Fr) (1791–1868).

Pyrogallic Acid 1786. *Disc.* by C. W. Scheele (Swed) (1742–86).

Pyrometer 1730. Petrus van Muschenbroeck (Hol) (1692–1761) *inv.* bar pyrometer. 1750. Josiah Wedgwood (1730–95) *inv.* silicious cones for pottery furnace temperature checking. 1782. Wedgwood *inv.* pyrometer for pottery furnaces. 1785. Jesse Ramsden (1735–1800) *inv.* pyrometer for checking temperature of survey rods. 1821. Prof. Daniell, of King's College, London *inv.* registering pyrometer. 1823. Princep *desc.* use of graduated series of alloys with known melting-points to act as pyrometers. 1823. Thomas Johann Seebeck *disc.* the way to generate electric current by keeping loops of dissimilar metals at differing temperatures. 1834. Jean-Charles Athanase, a watch-maker of Paris *disc.* the thermo-couple. 1850. Becquerel (Fr) *inv.* (*sic*) the thermo-couple. 1851. Ericsson *inv.* the air pyrometer. 1870. C. W. Siemens *inv.* platinum electric resistance pyrometer. 1887. Callender *inv.* (*sic*) platinum wire resistance pyrometer. 1891. Le Chatelier (Fr) *inv.* electric pyrometer. 1888. Thermo-electricity applied to pyrometry.

Pyrophone (sensitive flame, or singing flame) *Inv.* by Kœnig. 1869. "Lustre chantant" *inv.* by F. V. Kastner, of Paris.

Q

Quadrant *c.* A.D. 150. Ptolemy (100–160) used a stone solar quadrant with gradations marked on its polished side. Tycho Brahe (1546–1601) used large quadrant of his own design at Uraniberg Observatory, Hven, Denmark. 1672. Two reflecting quadrants *des.* by Sir Isaac Newton (1642–1727). 1725. George Graham (1673–1751) *inv.* mural quadrant erected at Greenwich observatory. 1730. Reflecting quadrant *inv.* by Thomas Godfrey (U.S.). 1750. Bird *inv.* quadrant used at Greenwich. (*See also* Sextant.)

Quandrature of the Circle (π) 1841. Determined to 208 places by Dr. Rutherford. 1846. Determined by Dase to 200 places, and by Dr. Clausen, of Dorpat, to 200 places. 1851. Determined by William Shanks of Houghton-le-Spring, Co. Durham to 527 places, and in 1853, to 607 places.

Quantum Theory 1900. Proposed by Max Planck (Ger). 1913. Planck's theory amplified by Neils Bohr (Dan).

Quaternions 1843. *Inv.* Sir Rowan W. Hamilton (1805–65). (*See also* Vector analysis.)

Quern 2500 B.C. Saddle quern used in Egypt. 1200 B.C. Rotary quern used in Syria. (*See also* Mill.)

Quaver (music) 1496. *Inv.* by Gafurius.

Quinine 1811. First obtained

from cinchona bark by Pfaff. 1820. *Disc.* (*sic*) by Pelletier and Caventou. 1838. Composition established by Baron von Liebig. 1865. Synthesized by W. L. Scott. 1908. Chemical structure deduced by Rabe. 1944. Synthesized by R. B. Woodward, of Boston, Mass., U.S.

R

Rabies *See* Viruses.

Radar (RAdio Direction And Range) 1888. Carl Hertz reflected and refracted centimetre-long radio waves. 1935. First experiments made near Daventry by Dr. Watson-Watt. 1937. *Const.* commenced on chain of 20 radar stations located from the Solent to the Firth of Tay. 1945. J. S. Hey used radar for detecting enemy V-2 rockets. "Gee," "Bloc," and "H2S" radar systems *dev.* by A. C. B. Lovell. (*See also* Radio-astronomy.)

Radiant Heat Properties of *disc.* by Macedonio Melloni (It) (1798–1854).

Radiator, Motor-car 1898. Honeycomb radiator (the steam condenser of 50 years earlier) first fitted to Cannstadt-Daimler motor-car, then on Mercedes and Albany cars. 1904. Louis Renault (Fr) *pat.* first rear-of-engine radiator. (*See also* Condenser, steam.)

Radio Telegraphy and Telephony 1895. Guglielmo Marconi (It) (1874—1919) sent first radio telegraphy signals in Italy. 1895. Alexander Popoff (Rus) and Eugene Ducretet (Fr) give practical demonstration at Leningrad. 1896. Marconi *pat.* apparatus in U.K. 1898. Radio telegraph link made from South Foreland lighthouse to South Goodwin light-vessel. 1900. Reginald Fessenden (U.S.) first transmits speech by radio. 1901. Prof. F. Braun *inv.* first crystal detector —the psilomelan. 1901 (Dec. 12). First trans-Atlantic radio link made from Poldhu, Cornwall to St. John's, Newfoundland. 1903. Arc Radio speech transmission *inv.* by Valdemar Poulsen (Dan). 1904. Sir Ambrose Fleming *inv.* twoelectrode valve (*q.v.*). 1905. Fessenden *pat.* his heterodyne system of detection. 1906. Lee de Forest (U.S.) *inv.* three-electrode valve. (The *inv.* of the feed-back valve circuit to amplify signals has been variously credited to Major Edwin Armstrong (U.S.), C. S. Franklin, Irving Langmuir, Meissner, and H. J. Round) Carborundum detector *inv.* by H. H. Dunwoody (U.S.); and G. W. Pickard *inv.* silicon detector. 1907. Fessenden broadcasts speech 100 miles by radio. 1924. Armstrong conceived idea of frequency modulation; 1933, perfected a system of static-free frequency modulation broadcasting which was tried in U.S. (First frequency-modulated transmission (broadcast) in U.K. not made until 1955.) 1927. Arnstrong *inv.* super-regenerative receiver. *c.* 1930. Armstrong *inv.* super-heterodyne receiver. 1947. J. A. Sargrove *inv.* printed radio receiver circuit. (*See also* Broadcast, radio.)

Radio-activity 1898. *Disc.* by Henri Becquerel (Fr). Induced radio-activity *disc.* by Frédéric and Irène Joliot-Curie (son-in-law and daughter, respectively, of Madame Curie).

Radio-astronomy 1931. Karl Jansky (U.S.) made first major study of radio atmospherics and found that their maxima occurred every 23 hours 56 minutes. 1937. Finding corroborated by Grote Reber (U.S.), who first suggested that they came from outer space. Skellet (1935), and Appleton (1939) *obs.* that radio waves were reflected from the ionosphere. (*See also* Radar and Radio telescope.)

Radiometer 1873. *Inv.* Sir William Crookes (1832–1919). **Radio Telescope** 1931. Karl Jansky (U.S.) made first radio telescope. 1937. Grote Reber (U.S.) made first dish-type radio telescope. (*See also* Radar and Radio-astronomy.)

Radium (element) 1898. *Disc.* by Pierre Curie (Fr) (1859–1906) and his wife Marie Sklodowska (Pol) (1867–1934). 1903. Sir William Ramsey and Frank Soddy *disc.* that radium slowly degenerated into helium. 1909. Radium *prod.* from uranium experimentally by Frederick Soddy. (1909. First gramme of pure radium bromide sold by Austrian Government for £10,000.)

Radon (element) 1900. *Disc.*

Rail-car 1899. Daimler petrol rail-car accommodating 18 passengers run on Wurtemburg Railways. 1903. First petrol rail-car run in G.B. on North Eastern Railway. 1913. First diesel-electric rail-car in use in Sweden.

Rail, Railway 1830. "T"-rail *des.* R. L. Stevens (U.S.). 1845. W. H. Barlow *inv.* "saddleback" rail.

Railway Levigated (Magnetic) *inv.* Louis Bachelet (Fr). **Railway Electric** 1837. Thomas Sturgeon propels a locomotive with electricity with the motor he *inv.* in 1832. 1838. Prof. Charles Page (U.S.), of Washington, D.C. *des.* an electric motor with which he successfully hauled a train between Washington and Bladensburg, on the Baltimore and Ohio Railroad in 1839. 1840. Uriah Clark, of Leicester built an electric locomotive which ran on a circular track at his home-town. Pinkus *pat.* electric railway with two extra conductor rails. 1842. Davenport (U.S.) runs an electric locomotive near Glasgow with the motor he *inv.* in 1837. 1847. Lilly and Colyon, of Pittsburg, Pen., U.S., *dev.* an electric railway. 1864. Bellet and De Rouvre *des.* electric railway with conductor rails. 1873. Hallez de Arros, of Nancy, *inv.* a battery-driven locomotive. 1884. E. M. Bentley and W. H. Knight *inv.* electric railway which operated in Cleveland, Ohio, U.S. (*See also* Locomotive, electric and Tramway, electric.)

Railway, Atmospheric 1824. Vacuum railway system *prop.* George Medhurst, 1812, *pat.* John Vallance.

Rain, Man-made 1945. First *prod.* by process *inv.* by V. J. Schafer (U.S.).

Rainbow Colours in rainbow explained by Harriot (1560–1621). (*See also* Roger Bacon and Vitellio.) True theory of rainbow explained by Antonio de Dominis (1566–1624). 1611. Theory further *dev.* by Johann Kepler (1571–1630). 1629. Theory perfected by René Descartes (1596–1650).

Rake, Horse *c.* 1807. Earliest record of use in Essex.

Ram, Hydraulic 1772. Ram with manually-operated valves *inv.* and set up at Oulton, Cheshire by John Whitehouse. 1796. J. M. Montgolfier (Fr) and Aimé Argand made ram automatic by placing a loose impulse-valve in its waste-pipe. 1797. Ram *imp.* by Matthew Boulton (1728–1809). Single-acting suction ram. *inv.* Bernoulli. 1858. Double-acting ram *inv.* by Nicholas Leblanc (Fr) (1742–1806).

Ram-jet Engine ("Athodyd,") or aero-thermo-dynamic duct engine. *See* Engine, jet.

RAMELLI, Agostino (It) (1531–90) 1588. One of the pioneers of technical handbook writing, with his book *Various and ingenious Machines* (*Le diverse et artificiose machines*).

RAMSDEN, Jesse (1735–1800) 1770. *Inv.* the first screw-cutting lathe and many other precision mechanical and optical devices.

RAMSAY, Sir William (1852–1916) Chemist and physicist who *disc.* many of the inert gases in the earth's atmosphere. (*See also* J. W. S. Rayleigh.)

RANKINE, William John Macquorn (Scot) (1820–72) *c.* 1860. evolved the scientific term "energy."

Rate-of-turn Indicator (aeronautical engineering) 1910. *Inv.* H. F. Wimperis. 1929. Schilovski and Cooke *inv.* improved instrument. 1930. Reid and Segrist *inv.* "Standard" R.A.F. instrument.

RAYLEIGH, John William Strutt, Baron (1842–1919) Chemist and physicist. Made efforts to discover a formula for the distribution of the emission of a black body over various wavelengths. Investigated the distribution of the inert gases argon, helium, krypton, neon and xenon. (*See also* Sir W. Ramsay.)

Rayon (cellulose acetate) 1855. Early rayon *inv.* George Audemars (Swit). 1884. Chardounet *pat.* spinneret process of *prod.* (*See* Chardounet.) 1890. First prepared by Schutzenberger (Ger). 1899. First filaments made by C. A. Bronnert (Ger). 1919. Stretch-spinning of rayon *inv.* by Edmund Thiele and Dr. Elsaesser (Ger) by cuprammonium process. 1930. Modern rayon *inv.*

Razor, Safety 1847. Safety razor with comb-teeth guard *inv.* William Samuel Henson (1805–88), of Chard, Somerset.

Reaping Machine 1799. Joseph Boyce *inv.* machine for cutting wheat and corn. 1800. Robert Mean *inv.* machine for cutting standing corn and grass. (*See also* mowers.) 1811. Salmon *desc.* reaping machine with clippers and delivery system. Smith, of Deanstown, Scotland, *desc.* reaping machine with rotary cutter and side delivery. David Cumming *desc.* rotary reaping machine. 1830. Budding *inv.* a small mowing machine. 1834. McCormick *pat.* his reaper. 1851. Marsh incorporated conveyer-belt in reaping machine. 1878. Appleby *inv.* knotting device for reaping machines. 1880. Appleby fits reaping machine with autobinder. 1936. First "baby" combine harvester *intro.*

REAUMUR, René Antoine Ferchault de (Fr) (1683–1757) 1730. *Inv.* thermometer with scale graduated 0° to 80°. (This scale now bears his name.)

Recoil, Gun 1867. Hydraulic recoil absorber *inv.* Sir William Siemens. 1898. French "75" gun with recoil absorber *intro.* (*See also* Gun.)

Recorders, Magnetic 1888. Oberlin Smith first *prop.* idea of magnetic sound storage. Valdemar Poulsen (Dan) *pat.* his "Telegraphone," or wire recorder. 1903. Wire recorders made in U.S.A. 1907. Directcurrent biasing principle *inv.* by Poulsen. 1920. Alternating-current biasing *inv.* by Poulsen. *c.* 1920. Kurt Stille (Ger) perfected the "Blattnerphone." *c.* 1920. Dr. Pfleumer (Aus) *inv.* plastic tape for magnetic recorders. 1921. Alternating bias current perfected by W. L. Carson and G. W. Carpenter (U.S.). 1927. J. A. O'Neill *pat.* paper tape in U.S. 1930. German "Magnetophone" tape in use. 1937. Plastic tape made and used in Germany. 1939. Cellulose acetate tape began to be used. *c.* 1940. Marvin Camras (U.S.) *imp.* the wire recorder.

Rectifier, Electrical 1902. Peter Cooper-Hewitt *intro.* mercury arc rectifier. 1924. First commercial form of selenium rectifier *intro.* 1926. Copper oxide rectifiers for low powers only *intro.* 1936. First pumpless, air-cooled, steel-clad mercury rectifiers installed in England.

Refining, Electrolytic 1869. Elkington's process of electrolytic refining instituted at Pembray, S. Wales. 1878. Holloway *inv.* first electrolytic process for refining copper.

Reflection (*See* Light.)

Reflex Action (biology) 1833. Term *intro.* by Dr. Marshall Hall (1790–1857).

Refraction (*See* Light.)

Refractometer 1874. First *inv.* and *const.* by Ernst Abbé.

Refrigeration 1000 B.C. Chinese poems by Shi Ching refer to ice cellars. A.D. 23–79. Pliny the Elder mentions new *inv.* of method of cooling drinks in snow, stored and insulated by grass. Indians, Egyptians and Estonians chilled water by placing it in shallow, porous pans and leaving them out-of-doors at night. 1842. H. Benjamin froze food by immersion in mixture of ice and brine. 1852. Lord Kelvin and Prof. Rankine *prop.* first open-cycle refrigeration.

Refrigerator 1834. Jacob Perkin *inv.* first vapour-compression refrigerator. 1845. Dr. John Gorrie (U.S.) *inv.* air refrigerator. 1851. Filed first U.S. patent for making ice by compressed air. 1857. James Harrison (Australia), of Geelong, used Perkins's refrigerator to export meat by ship. 1860. Carré (Fr) *inv.* ammonia absorption refrigerator. Carbon dioxide used as a refrigerant. *c.* 1860. Ether, methyl chloride, and sulphurous acid used as refrigerants. 1862. Gorrie's refrigerator *imp.* by Kirk, who added a regenerator. 1873. Kelvin-Rankine refrigerator made in practical form by Giffard and by Coleman and Bell. 1876. Dr. Carl Lindé *intro.* ammonia refrigerator. 1905. Platen and Munters (Swed) *inv.* absorption-type ("Electrolux") R. Philips (Hol) *prod.* refrigerator for making liquid air at atmospheric pressure. 1930. T. G. N. Haldane had installed at his home the first heat-pump of his own design. *See also* Cryogenics.

REGNAULT, Henri Victor (Fr) (1810–78) Physicist who pioneered accurate measurement of heat and determined

pressure of saturated water and heat of vaporization of water at various temperatures.

REISS, Philip 1861. *Inv.* membrane microphone.

Relativity, Theory of 1905. First *prop.* by Swiss Patent Office clerk Albert Einstein (1879–1955). 1915. Theory established by R. A. Millikan (U.S.).

Relay, Electric 1837. First devised by Edmund Davy and termed a "renewer."

Repeating Circle (surveying) 1787. *Inv.* Étienne Lenoir (Fr).

Reserpine (drug) 1956. Synthesized by R. B. Woodward, of Boston, U.S.

Resins, Synthetic 1901. Alkyd synthetic resin *disc.* by William Smith. 1918. Synthetic resin made from ammonia and carbon dioxide—synthetic urea made, which, plus formaldehyde *prod.* "Plaskon," "Beetle," and "Scarab" wares. 1926. Commercial alkyd resins derived from glycerine and napthaline products.

Respirator *c.* 1853. Charcoal respirators and air filter *inv.* Dr. John Stenhouse. 1870. Firemen's respirator *inv.* Prof. John Tyndall (1820–93). 1879. Dr. Woillez (Fr) anticipated the idea of a respirator. 1929. Phillip Drinker, of Harvard, U.S. *inv.* the "Iron Lung." (A respirator was also devised by Dr. Marshall-Hall, of Edinburgh.)

Respiration, "Cheyne-Stokes" First *desc.* by William Stokes of Dublin (1804–78), and John Cheyne, a Scottish doctor practising in Dublin.

Retort (coke-oven) 1796. Beehive-type gas-retort *inv.* 1888. Inclined coke-oven *inv.* by Coze, of Rheims. 1899. Köppers *inv.* coke-oven. 1900. Vertical coke-oven *inv.* by Bué. (The horizontal gas-retort *inv.* by Coppée (Bel) and *dev.* later by Hüssener and Carl Otto.)

Revolver 1835. Col. Samuel Colt, of U.S. Army *inv.* six-chambered revolver.

Reynaud's Disease 1862. First *desc* by Reynaud.

Rhenium (element) 1925. First identified.

Rhodium (element) 1803. *Disc.* by William Hyde Woolaston (1766–1828).

Rifle 1498. Rifled hand-guns used at shooting-match at Leipsig. 1568. Straight-grooved rifles used in Germany. 1563. Spirally rifled hand-guns referred to in Swiss law. 1645. Bavarian rifle regiments known to have existed. 1827. J. N. Dreyse *inv.* the needle-rifle, adopted by Prussian army in 1842. 1833. Minié rifle *inv.* by Minié, of Vicennes, and adopted by French army—by British army in 1852. 1836. Saw the advent of the Brunswick Rifle: 1853, that of the Enfield short rifle; 1860, that of the Enfield long rifle; 1864, the Schneider; 1866, the Chassepot; and 1871, the Martini-Henry.

Rifle, Sporting 1856. Needham *inv.* sporting rifle. (Other early English sporting rifles *inv.* by Green, Davies and Murcott.) 1875. Anson and Deeley *inv.* sporting rifle. (Other *invs.* include those of Purdey and Greener (ejector sporting rifle).)

Riveting Machine *c.* 1805. Mark Isambard Brunel established army boot factory with riveting machine for riveting soles. (Hand work was reverted to as soon as war was over.) 1838. Sir William Fairbairn *inv.* riveting machine to cope with boilermakers' strike at his works.

Roads *c.* 1785. Waterbound flint ("Macadam") roads *inv.* by John Loudon Macadam (1756–1836). 1815. First "Macadam" road laid in England at Bristol by Macadam and R. L. Edgeworth. 1830. First tarmac road in England laid in Nottinghamshire. 1832. Asphalt road first *dev.* in France by Sassenay. 1854. Compressed rock and asphalt road laid in Paris. 1865. Concrete roads *intro.* into U.K. in Scotland. 1870. Compressed rock and asphalt roads *intro.* into U.S. 1870–90. De Smedt and Richardson *intro.* watertight chip-sand-bitumen roads. 1873. Tarmac roads *intro.* into U.S. 1879. Road-making machines *intro.* in Germany. 1892. Concrete roads *intro.* into U.S. 1901. Guglielminetti (It) *intro.* hot bitumen treatment of De Smedt's roads. Three-layer road-making system *intro.* by P. M. Tréseguet (Fr). (*See also* Paving, road.)

Road Vehicle, Steam (passenger and freight-carrying) 1763. Nicholas Joseph Cugnot (Fr) (1725–1804) built a model steam road vehicle. 1769. Cugnot first steam road vehicle which reached a speed of $2\frac{1}{4}$ mph. 1780. Charles Dallery (Fr) *des.* and *const.* a steam road vehicle with a multi-tubular boiler. 1784–86. William Murdock (1754–1839) *des.* and *const.* a model steam road vehicle. 1786. William Symington (Scot) *des.* a steam road vehicle model with rack transmission-gear. 1788. Robert Fourness and James Ashworth, of Halifax, Yorkshire, *pat.* gear-drive, steam road vehicle with a three-cylinder engine run on high-pressure steam. 1788. Nathan Read (U.S.) *des.* and *const.* model steam-carriage with a multi-tubular boiler. 1801. Richard Trevithick, of Cornwall, *const.* a steam road vehicle which ran at Camborne. 1802. Trevithick's steam road vehicle *pat.* 1803. Trevithick's vehicle ran in streets of London. 1805. Oliver Evans (U.S.) (1755–1819) *des.* an amphibious steam vehicle—the first steam road vehicle to run on roads of U.S.A. 1815. Josef Bozek, of Prague *des.* a steam road vehicle to carry three persons. 1819. George Medhurst (1759–1827) *des.* a steam road vehicle which ran between Paddington and Islington, in London. 1821. Julius Griffith, of Middlesex, *des.* steam road vehicle with change-speed gears and spring-mounted engine. 1824. David Gordon *inv.* steam road vehicle worked by backward-thrusting legs. (Idea also tried by Sir Goldworthy Gurney (1793–1876) the following year.) 1824. W. H. James, of Birmingham and Sir James Anderson, of Ireland, *des.* steam road vehicle with two twin-cylinder engines each of which drove one rear wheel. 1825. Timothy Burstall and John Hill (Scot) *des.* four-

wheel-drive steam road vehicle with ratchet-type differential drive, and of great constructional ingenuity. 1828. James and Anderson *des.* 18-seater steam road vehicle capable of 15 m.p.h. *Other outstanding steam road vehicle designs are the following:* Sir G. Gurney, 1823, 1827, 1831. Walter Hancock (1799–1852), 1829, 1830, 1831, 1832, 1833, 1834, 1835 and 1836. John Scott Russel, of Edinburgh, 1834 (ran between London and Kew). Francis Macerone and John Squire, 1833, 1834, 1843. Frank Hills, of Deptford, 1839–43. (His last steam road vehicle made long runs from London to Brighton, Windsor and Hastings.) *Among other less important British steam road vehicle builders were:* Nasmyth, Heaton, Battin, Gibbs, Hanson, Roberts, Gough, Redmund, Napier, Ogle and Summers, Church, Maudslay Palmer, Fraser and Alexander Gordon; and in France, Dietz, Omont, Cazalet and Pecquer. *From 1861 onwards the following are the most important invs. of steam road vehicles:* 1861, W. O. Carrot, of Leeds; 1862, Yarrow and Hilditch, of Greenwich; 1862, Tangye, of Birmingham; 1862, A. Patterson; 1867, H. P. Holt; 1867, R. W. Thomson; 1868, R. E. B. Crompton; 1868, Catley and Ayres, of York; 1869, L. J. Todd, of Leith; J. H. Knight, of Farnham, Surrey; 1868, Joseph Ravel (Fr); 1869, Andrew Nairn, of Leith; 1874, Charles Randolph; H. A. C. Mackenzie, of Diss, Norfolk; 1877, A. B. Blackburn (first use of liquid fuel); 1880, Sir Thomas Parkyn and A. H. Bateman; 1881, J. C. Inshaw; 1882, Copeland (U.S.), of Philadelphia (steam tricycle); 1873, Amédée Bollée, Snr., *des.* and *const.* 12-seat steam road vehicle (also *des.* steam road vehicles in 1878, 1879, 1880 and 1881).1883–97, Count Albert De Dion, in partnership with Georges Bouton and Charles Trepardoux made successful steam tricycles and road coaches. 1876, 1887, 1889, 1890, 1897, 1902, 1905, Léon Serpollet (Fr) *des.* and built light steam cycles and carriages, winning the land speed record in 1902 at 75.06 m.p.h. 1894, 1897, Georges Scott, of Epernay (Fr), *des.* steam wagonette and steam omnibus. 1862. J. I. Thornycroft. 1899. First steam motorcar *des.* by Stanley Brothers, of Newton, Massachusetts. 1900. First steam car by White Brothers, of Cleveland, Ohio. 1903. First Doble steam car appeared. (Production of *imp.* models continued until 1932; when car achieved a speed of 95 m.p.h.) (*See also* Traction-engine, steam.)

Rocket 1232. First use in war by Chinese against Mongols at sieges of Kiai-fung-fu and Ho-yang. 1258. Rockets mentioned at Cologne. 1282. "Flying-fire-arrow" *inv.* by Wei Chiang (Ch). 1379. Used by Paduans at battle of Chiozza and by Venetians the following year. 1688. Germans experiment with rockets weighing 1 cwt. 1802. Col. William Congreve *des.* life-saving rocket.

1806. Boulogne bombarded with flame-carrying rockets fired from ships. 1807. Rockets used at Copenhagen; 1809, in the Basque Roads; 1812, at battle of Blankenberg; and 1813, at Leipsig. 1914–18. Le Prieur (Fr) *des.* rocket which could be fired from aircraft. 1934. Germans successfully fired rockets to height of $1\frac{1}{2}$ miles. 1939 (Oct.). German rocket climbed to height of 5 miles with complete stability. (*Des.* by team led by von Braun at Peenemünde.) 1943 (Oct.). First "V-2" rocket rose beyond earth's atmosphere and travelled 125 miles on a fuel of alcohol, liquid oxygen and hydrogen peroxide. 1944 (Sept. 8). First "V-2" rockets fell at Chiswick and Epping, London. 1957 (Oct.-Nov.). Russian "sputniks" I and II successfully placed in orbit by rockets. 1959 (Sept. 13). First rocket to land on Moon—Russian "Lunik II." (*See* Rocket-propelled aircraft.)

Rocket-propelled Car 1928. First made by Franz Opel (Ger) to *des.* of Max Valier (Ger). 1930. Valier started using liquid fuel of petrol and oxygen.

Rocket, Life-saving 1791. John Bell propelled 8 in. shell with chain attached. 1802. William Congreve *des.* life-saving rocket. 1807. Capt. G. W. Manby *inv.* mortar-type rocket. 1823. Hase, of Saxthorp, Norfolk, *inv.* skeleton rope-reel for rocket line. Henry Trengrouse, of Helston, Cornwall, *inv.* life-saving rocket used at wreck of H.M.S. *Anson.*

ROEBUCK, Dr. John (1718–94), of Sheffield. 1746. *Inv.* lead-chamber process for making sulphuric acid commercially.

Roller, Agricultural 1841. *Inv.* by William Crosskill for rolling grass. (He also *inv.* clod-crushing roller.)

Roller, Steam (road) 1738. Polonceau (Fr) *inv.* road-roller filled with water or crushed stone. 1787. Hand-rammers displaced by Cessart's (Fr) *inv.* of iron hand- or horse-drawn roller. 1861. Lemoine (Fr) *inv.* steam road-roller. 1862. Ballaison (Fr) *inv.* steam road-roller with two equally sized rollers. 1863. Clark and Butler *des.* three-wheeled road-roller. 1867. Aveling, of Maidstone, Kent, *des.* steam road-roller. 1902. First internal-combustion-engined road-roller used in England.

Rolling-stock, Railway 1869. First articulated vehicle *desc.* by T. Claxton-Fidler.

Rolls, Metal 1784. *Inv.* by Cort. (*See* Iron.)

RÖNTGEN, Wilhelm Konrad (Ger) (1845–1923) 1895. *Disc.* X-rays.

Roof, Mansard *c.* 1650 *Inv.* by François Mansard (Fr) (1598–1666).

Roof, Motor-car Sliding 1929. *Inv.* by T. P. Colledge.

Rope, 1792. Flat rope *inv.* by John Carr.

Rope-making Machine *Dev.* by Leonardo da Vinci (1452–1519). 1754. *Pat.* by Richard Marsh. 1783. *Inv.* by Sylvester. 1792. *Pat.* by Rev. Edmund Cartwright.

Rope, Wire 1832. First made.

1840. Robert Sterling Newell takes out first *pat.* for making wire rope.

Ropeway, Aerial 1441. First on record built at Danzig by Adam Wybe, of Hallingen, Holland. 1857. Henry Robinson, of Seattle, Yorkshire, granted first U.K. patent for aerial ropeway. 1868. John Buddle *intro.* endless-rope mine haulage into U.K. Charles Hodgson *inv.* endless aerial ropeway system, and later the fixed carrying-rope system *imp.* in 1873 by Adolf Bleichert (Ger). (Later *imp.* by Theodor Otto, W. T. H. Carrington, Kremer and Theobald Obach.) (*See also* Power transmission and Telpherage.)

Rosetta Stone 1799. *Disc.* (inscriptions on rosetta stone deciphered by Dr. Thomas Young, of Taunton, Somerset.

Rosin 1912. First combined with synthetic resins (*q.v.*) by K. Albert.

Rotifers (animaculae) First *disc.* by Antonie van Leeuwenhoek (Hol) (1632–1723).

ROUTH, E. J. (1831–1907) 1878. *Prop.* theory of automatic control equations.

Rubber 1730. Rubber (caouchouc) *intro.* England from Brazil. 1736. Charles M. de la Condamine (Fr) *intro.* rubber into Europe (*sic*). 1770. "India-rubber" first mentioned for erasing pencil-marks by Joseph Priestley. 1820. Mastication of rubber *disc.* by Thomas Hancock. 1839. Vulcanization process *inv.* by Charles Goodrich, of New Haven, N.Y. 1843. Word "Vulcanization" coined by Thomas Hancock's friend

Brockedon. Goloshes made from rubber. 1855. First attempt to cultivate rubber plant in East and West Indies. *c.* 1870. Rubber came into common use as electrical insulator. 1873. First rubber plant seedlings sent from Brazil to Calcutta, and thence to Ceylon, Singapore and Java. 1877. 2,000 seedlings brought to Kew Gardens, London, and sent thence to British Guiana, Honduras and the West Indies. 1894. Rubber production abroad effectively started.

Rubber, Synthetic 1860. C. G. Williams (1829–1910) *prod.* isoprene. 1910. S. V. Lebeder (Rus) makes synthetic rubber from butadeine. 1924. J. C. Patrick (U.S.) makes *dev.* in synthetic rubber *prod.* from ethylene dichloride and sulphide of potassium. 1930. "Buna" synthetic rubber (from BUtadeine and NAtron) manufactured in Germany. 1931. Neoprene synthetic rubber *intro.* 1940. Neoprene made from acetylene gas in U.S., and in U.K. from 1960. W. A. Tilden (U.S.) (1842–1926) *disc.* isoprene could be turned into synthetic rubber.

Rubidium (element) 1861. *Disc.* by R. W. Bunsen (1811–99).

RÜHMKORFF, Heinrich Daniel (Ger) (1803–77) 1851. *Dev.* transformer *inv.* by M. Faraday and J. Henry into the induction-coil. (Grafton Page (U.S.) built an induction-coil in 1838.) (*See also* Coil, induction.)

Rule, Slide Circular slide rule *inv.* William Oughtred (1575–1660).

Rule, Log Slide *Inv.* English Mathematician William Oughtred.

Ruler A.D. 79. Bronze ruler found at Pompeii. 1723. Parallel ruler *inv.* by Byon. 1771. Rolling ruler *inv.* by Eckhardt.

Ruling Machine 1782. *Inv.* by a Dutchman living in London. *c.* 1803. *Imp.* by Woodmason, Payne, and Brown.

Rum 1530. First *prod.* in Jamaica.

RUMFORD, Count (*See* Thompson, Benjamin.)

Ruthenium (element) 1845. *Disc.* by Klaus.

RUTHERFORD, Sir (later Lord) Ernest (1871–1937) (Knighted, 1914 and created a baron, 1931; *b.* Nelson, New Zealand) Pioneered nuclear physics with Frederick Soddy and others.

RYDBERG (Swed) *Disc.* reciprocal formula for spectroscopic lines of wave-length series.

S

SABINE, Wallace Clement (U.S.) (1868–1919) *c.* 1900. Pioneered investigations into the acoustics of theatres, lecture-rooms, etc.

Saiger (metallurgical process) 12th cent. *Inv.* by a Venetian. 16th cent. Process in common use.

SALINAS, Franciscus (Sp) (1513–90) 1577. Pioneered the development of German organ-builder Arnolt Schlick's musical mean-tone system.

Salicylic Acid 1899. First synthesized, and *intro.* as aspirin.

Salk Anti-polio Vaccine 1952. First *dev.* by Dr. Jonas Salk (U.S.). 1955. First mass use of his vaccine.

SALLO, Denys de (Fr) (1626–69) 1665. *Pub.* the first scientific "review"—*Journal de savants*.

Salvarsan ("606") 1909. *Disc.* as a remedy for syphilis by Paul Ehrlich (1854–1919), of Frankfurt. 1912. Ehrlich synthesized neo-salvarsan and *intro.* its use.

Samarium (element) 1879. *Disc.* by Lecoq de Boisbaudran (Fr).

Sand-blast 1871. Compressed-air sand-blast *inv.* by B. C. Tilghman (U.S.).

Sand, Pressing Flowers in 1633. Process *disc.*

Sarrusophone (musical instrument) 1856. *Inv.* by Rène-Louis Gautrot, of Paris.

Saturn, Moons of 1659. First moon *disc.* by C. Huygens (1629–95). Eighth moon *disc.* by William Russel and W. C. Bond. 1850. W. C. Bond and William Rutter Dawes (1799–1868) *disc.* Saturn's "crepe ring."

SAUSSURE, Horace-Benedict de (Fr) (1740–99) 1783. *Inv.* the hair hygrometer.

SAUSSURE, Nicholas Théo de (Fr) (1767–1845) 1804. *Disc.* that saltpetre (potassium nitrate) promoted the growth of many plants and cereals.

SAUVEUR, Joseph (Fr) (1653–1716) Devised method of sound frequency measurement and coined the words "noeud" ("node") and "ventre" ("belly") as applied to sound-waves.

SAVART, Felix (Fr) (1791–1841) *c.* 1820. Determined the limits of audibility of the normal human ear, making use of a siren and cog-wheel of his own invention.

SAVERY, Thomas (1650–1715) Pioneered early *dev.* of the steam-engine. 1698. *Pat.* 1 h.p. engine known as the "Miners' Friend." 1712. Installed such an engine at Camden House, Kensington, London.

Saw, Band 1808. Ineffectual steel band saw *pat.* by William Newbury. 1843. *Imp.* by J. G. Bodmer. 1846. Band saw *pat.*

by Exall, of Reading. 1849. Re-*inv.* by Lemuel Hedge (U.S.). 1855. First practical band saw *inv.* in France by Perrin. (Band-saws were also *pat.* by M. I. Brunel and R. Stephenson.)

Saw, Circular 1790. *Inv.* (*intro.* into England) by Walter Taylor, Southampton. 1791–93. *Inv.* by Gen. Sir Samuel Brunswick, of Maine, U.S. 1807. M. I. Brunel *inv.* and used circular saws of many sections. 1824. *Imp.* wide teeth *inv.* Robert Eastman. 1824. George Smart, of Westminster used first circular saw in London. *c.* 1824. Bentham *inv.* the slotted bench, guide, and fence. 1824. Multiple circular saws *inv.* by Sayers and Greenwood.

Saw, Fret 1866. Vibrating fret-saw *inv.* by James Kennan, of Dublin.

Saw, Hand Pre-7th cent. B.C. Assyrians used iron saws. Paintings of saws found at Herculaneum. Palladius *desc.* non-framed hand saws. Cicero, in an oration of Cluentius mentions wood saws. Egyptian saws found in Thebes tombs, with paintings of such saws in use. Leonardo da Vinci devised a hand-saw for marble-cutting.

Sawing Machine, Veneer 1824. *Inv.* by Alexander Craig (Scot).

Saw-mill 4th cent. Ausonius *desc.* water-driven saw-mill on River Moselle used for cutting marble. Set up in the following places: 1322, at Augsberg; 1420, Madiera; 1427, Breslau; 1490, Erfurt; 1530, in Norway; 1540, Holstein; and Joachims-thal; 1555, Lyons; 1556, Saar-dam, Holland. 17th cent. *Intro.* into England, the first ones being destroyed by sawyers in 1663.

SAX, Adolph 1840. *Inv.* the saxophone.

Scales (*See* Balance.)

Scandium (element) 1869. Predicted by Mendeléev. 1879. *Disc.* by L. F. Nilson (Swed).

SCHEELE, Carl Wilhelm (Ger) (1742–86) Outstanding chemist of 18th cent. *Disc.* chlorine, glycerine, milk-sugar, and hydrofluoric, lactic, oxalic, citric, tartaric, uric, and arsenic acids; together with many other compounds. 1783. With Torbern Bergman (1735–84) *disc.* molybdenum.

SCHEINER, Christopher (1575–1650) 1611. One of the first to *obs.* sunspots through a telescope; Johannes Fabricus having seen them the previous year.

SCHLICK, Arnolt (Ger) (*c.* 1460–1517) 1511. Proposed the mean-tone musical system.

SCHMIDT, Wilhelm (Ger) (1858–1924) 1900. *Inv.* the locomotive superheater.

SCHÖNBEIN, C. F. (Swit) (1799–1863) 1840. *Disc.* ozone. 1846. *Disc.* gun-cotton.

Schrödinger Equation (mathematics) 1926. Evolved by Edwin Schrödinger, of Vienna.

SCHÜTZENBERGER, Paul (Ger) (1829–97) 1865. *Inv.* acetate-rayon called "Celanese," which was first *prod.* commercially in 1904.

SCHWABE, Heinrich Samuel (Ger) (1789–1875) 1851. *Disc.* periodicy of sunspot maxima.

SCHWANHARDT, George (Ger) (1601–67) Pioneered technique of jewel-cutting to glassware.

SCHWEIGGER, Johann Salomo Christoph (Ger) (1779–1857) 1820. Increased effect of current in Oersted's experimental "ampmeter" by coiling wire round a magnetized needle. (Effect further *imp.* by Leopoldo Nobili (It).)

Scissors, Mayo (surgical instrument) *Inv.* by Brothers William James (1861–1939) and Charles Horace (1865–1939) Mayo curved surgical scissors *inv.* by Marc Antoine Louis, a Parisian surgeon.

Scott's Dressing (surgical) *Inv.* John Scott (1799–1846).

Scraper, Earth 1783. *Inv.* William Driver (*pat.* No. 1366).

Screw *Inv.* credited to Archimedes of Syracuse (*c.* 287–212 B.C.).

Screw, Metal *c.* 1405. Mentioned by Kyeser in his *Bellafortis*. 1480. Found in clock *const.* and key-smithing.

Screw, Wood 1845. Woodscrew-making machine *pat.* by Japy (Fr). Japy's *pat.* was *imp.* by T. J. Sloan, of New York and later employed by William Angel, of Providence, Rhode Island. Sloan's *pat.* later sold to Nettlefold, of England.

Screw-cutting (making) Machine Leonardo da Vinci (1452–1519) used two male threads as masters to cut a thread, and explained how a thread could be varied in pitch by gear ratio variation. 1798. David Wilkinson (U.S.) *inv.* nut and bolt making machine. 1836. Die-holder and dies *inv.*

James Tracey, of Pembroke, Wales. Screw-cutter *inv.* N. S. Heineken, of Sidmouth. 1842. Nut and bolt finishing machine *inv.* Micah Rugg (U.S.). 1861. Spencer (U.S.) *inv.* automatic "brain-wheel" screw-cutting machine. 1879. C. W. Parker *inv.* first British automatic screw-making machine. 1895. Multi-spindle lathe *inv.* in U.S. 1900. Multi-spindle lathe *inv.* in Sweden.

Screw-thread 1841. Standard thread *inv.* by Sir Joseph Whitworth. *c.* 1860. Whitworth's standard thread with a 55° angle became standardized in England. 1864. Sellers, president of the Franklin Institute, *inv.* the American standard thread, which, in 1868 was officially adopted in the U.S.

Scythe 1791. Abraham Hill *inv.* steel-edged, iron-backed scythe. 1795. Arnold Wilde *inv.* scythes, sickles and hay-knives of fused iron with steel edges welded thereon.

SECCHI, Angelo (It) (1818–78) With H. C. Vogel, founded the science of stellar spectroscopy.

SEEBECK, Thomas Johann (Ger) (1770–1831) 1823. *Disc.* the thermo-electric couple. (*See* Peltier, J. C. A.)

SEABORG, Glen T. (U.S.) 1952–53. With A. G. Liorso (U.S.) *disc.* the elements Berkelium, Californium, Einsteinium, Fermium and Mendelevium.

Secretin (anatomy) 1902. *Disc.* by William Maddock Bayliss (1860–1924) (and Starling).

Sedative Powder 1732. Dover sedative powder *inv.* by Capt.

Thomas Dover (1660–1742), of Warwick.

SEGUIN, Edouard (Fr) (1812–1880) Pioneered early locomotive *des.* and *const.* in France.

Selenium (element) 1817. *Disc.* in riolite by J. J. Berzelius.

Selenium Cell 1873. Electrical resistance of selenium affected by light *disc.* by Willoughby Smith. 1875. Selenium cell for conversion of light into electricity *dev.* by G. R. Carey. (*See also* Photo-electric cell.)

Self-starter (automobile) 1912. A. H. Midgley *pat.* electric self-starter with gear-ring on engine flywheel. 1912. V. Bendix *inv.* automatic gear-coupling device for self-starters.

Semi-conductors (transistors) 1834. Munk (Munck) Rosenschöld *disc.* asymmetric electrical conduction in some solids. 1874. F. Braun re-*disc.* Rosenschöld's findings when examining metal rectifiers. Semi-conductivity of certain sulphides demonstrated. 1906. Semi-conductors first used. Carburendum radio detector *inv.* by Henry Harrison Dunswoody (U.S.) (1842–1940). "Catwhisker" radio detector *inv.* by Greenleaf Whittier Pickard (U.S.) (*b.* 1877). 1911. Oscillating diodes similar to "tunnel diodes," began to be used. 1915. Rectifying action of germanium *disc.* by Carl Axel Frederick Benedicks (Swed). 1924. Losser *disc.* that many crystals rectified electric current. 1925. Details *pub.* of what appears to be an n.p.m. transistor in Canada. 1930. R. S. Ohl experiments with silicon rectifiers.

1942. High-back-voltage germanium rectification *disc.* by Seymour Benzer at Purdue University, but not published for two years on account of the war. 1948. First public announcement of *inv.* of germanium rectifiers.

Separators (*See* Centrifuges.)

Serpent (musical instrument) *Inv.* by Edmè Guillaume, of Auxerre, France, as a modified German "Grosszinc."

SERPOLLET, Léon (Fr) Pioneer of steam road vehicles of late 19th cent. in France. 1887–88. *Inv.* steam tricycle. 1902. One of his vehicles held world speed record of 75.06 m.p.h. 1903. Gardner-Serpollet steam car *intro.* into England.

SÉNARMONT, Henri Hureau de (Fr) (1808–62) Mineralogist who, in 1847, demonstrated that the thermal conductivity of anistropic crystals depends on direction of heat-flow.

SENEFELDER, Alois (1771–1834) 1796. *Disc.* art of lithography.

Serum, Anti-diptheria 1890. Anti-toxin *disc.* Emil von Behring (Ger) (1854–1917). Serum *inv.* by Emile Rouse (Fr) (1853–1933). (*See also* Bacilli.)

SERVITUS, Miguel (Sp) (1511–53) 1553. *Pub.* first clear account of the pulmonary circulation of the blood. 1555. His theory seconded by Realdo Columbo of Padua (1516–59). (Servitus was burned at the stake.)

Servo-control 1867. John McFarlane Gray (1832–93) *des.* steam steering-gear for S.S. *Great Eastern*; probably the first application of the hydro-servo, or "follow-up" principle. 1870. A. B. Brown, of Edinburgh, *pat.* servo-motor for hydraulic steering by steam, air, or oil. 1872. Joseph Farcot (Fr) *inv.* feedback linkage for ship steering-gear. 1878. E. J. Routh (1831–1907) provided equations for solving the general problem of auto-control devices. 1882. P. W. Willans, of Rugby, *inv.* two-stage solenoid-operated valve to steam-electric generator. 1884. Proell (Ger) *inv.* practical hydro-relay governor.

Sewing Machine 1790. Matthew Saint *pat.* first sewing machine (unsuccessful). 1829. French tailor Thimmonier, of Amplepuis, Rhône, *pat.* unsuccessful chain-stitch machine. 1834. Walter Hunt's *pat.* 1841. Elias Howe (U.S.) *pat.* successful machine. 1850. Various *pats.* by Welhu and Wilson, Grove and Baker, and Caillehaut. 1851. Domestic sewing machine devised as a lock-stitch machine by Isaac M. Singer, of New York. 1852. Otis Avery's *pat.* 1854. Journau Leblond's *pat.* 1860. Two-needle machine *inv.* by W. N. Wilson, of London. 1860. Two-thread machine *inv.* by George Wright, of Ipswich.

Sextant 1550. *Inv.* by Tycho Brahe (1546–1601). 1699. *imp.*

Shaduf (counterbalanced baler) *c.* 2000 B.C. Represented on an Akkadian seal. *c.* 1400 B.C. In use in Egypt for irrigation purposes.

Shaping-Machine 1836. James Naysmyth (1808–90) *inv.* shaping-machine as distinct from M. I. Brunel's of *c.* 1805, which

was used at Portsmouth Dockyard for shaping ship's blocks.

Shield, Tunnelling 1890. *Inv.* and *intro.* by J. H. Greathead. (M. I. Brunel also *inv.* a three-stage rectangular tunnelling shield.

SHILLIBEER, George 1829. *Des.* London's first horse-bus which started to run on July 4, and carried 22 passengers, using three horses. (*See* Omnibus, Horse.)

Ship Pre-3000 B.C. Sailing ships in use. A.D. 370. Man-powered paddle-ship probably *inv.* by the unknown author of *De Rebus Bellicus.* 1436. Ox-powered ship with three sets of paddles *inv. c.* 1490. Leonardo da Vinci *desc.* paddle-driven ship. *c.* 1500. Descharges, of Brest, *inv.* portholes. 1543. Blasgo de Garay (Sp) *des.* man-powered paddle-ship. 1558. Julius Scaliger *desc.* another similarly powered ship. 1588. Augustin Ramelli *desc.* man-powered paddle-ship. 1595. Pieter J. Livorn, Hoorn, (Hol), *inv.* shallow-draught slender, three-masted "fluitschip." 1597. Roger Bacon *desc.* man-powered paddle-ship. 1618. David Ramsey and Thomas Wild-goose *pat.* a steam-driven ship. 1661. The Marquis of Worcester *pat.* ship with current-driven paddle-wheels which wound the vessel along a rope fixed to land. 1661. Toogood and Hayes *pat.* vessel propelled by sucking in water at the bow and forcing it out at the stern. 1664. Sir William Petty *inv.* double-bottomed ship which sailed from Liverpool to Holyhead and back in July.

1682. Horse-driven paddle-ship used at Chatham. 1690. Denis Papin (Fr) (1647–1712) *des.* ship powered by a gunpowder engine. 1696. Shipbuilding first made a science by Hoste. 1702. Capt. Thomas Savery (1650–1715) proposed a ship driven by steam. 1707. Papin suggested a ship driven by steam. 1709. Conrad used wind-force to mechanically propel a ship. 1730. Dr. John Allen *inv.* a mechanically propelled ship. 1737. Jonathan Hills *des.* first steamship. 1760. William Henry (U.S.) tried a steam launch on the Conegosta River. 1770 (summer). Compte d' Auxiren and Compte Follenay built a steamship which sank in the River Seine. 1773. Maillard *inv.* a clockwork-driven ship. 1775. Steamship with engine having 8 in. cylinder unsuccessfully tried on River Seine. 1775. Fitch and Henry (U.S.) *des.* steamship. 1777. Compte d'Auxiren built a ship with "fire-driven wheels" and tried it on River Danube, racing for 1,000 livres against seven new boats *inv.* by a Venetian. 1783. Marquis Jouffroy d'Abbans *des.* the 120 ton paddle ship *Pyroscaphe* and ascended the River Seine for 15 minutes. 1783. Fitch (U.S.) sailed steamship on Delaware River, and Rumsey, of Virginia, one on the Potomac River. 1785. Oliver Evans (U.S.) *des.* and built amphibian which plied on Schuyllkill River. 1787. John Wilkinson, of Bradley Forge *des.* and built 70 ft. canal-boat of iron, launching it on the River Severn. 1787. Experi-

ments made in U.S. with steam-pumped water from bow to stern as propulsion. 1797. Harriot *inv.* ship driven by falling water. 1801. Symington (Scot) *des.* and builds steamship *Charlotte Dundas*. 1801. Tresmere *inv.* a weight-driven ship. 1802. Patrick Miller equips two paddle-boats with Symington's engines. 1807. Robert Fulton (U.S.) sailed steamship *Clermont* from New York to Albany. 1808. 84 ft. iron ship *Manchester* built. Iron ship *Alburka des.* and built for Lander's Niger Exploration. John Stevens (U.S.) sailed steamship *Phoenix* from New York to Delaware. 1810. Joseph Hardy, of Rathmines Island *inv.* paddle ship which plied the canal at Portobello at 3 m.p.h. 1812. First successful steamship in Europe—*Comet* built at Glasgow by Henry Bell, and started on regular service from Glasgow to Greenock on the Clyde. 1815. Jevons, of Liverpool, sailed iron-built ship on River Mersey. 1818. First ocean-going steamship—*Savannah* launched (part paddle, part sails). 1821. First iron, ocean-going steamship *des.* with an oscillating-cylinder engine by Aaron Manby and named after him. (Built by Horsley and Co., and sent to London, where it was assembled before sailing to Paris under command of Sir Charles Napier (1822).) 1827. Dutch ship *Curaçao* made all-steam voyage Rotterdam to West Indies. 1827. Sir William Congreve propelled ship by capillary attraction of glass plates or sponge-wheels instead of paddle-wheels. 1829. 74-gun ship sheathed in rubber built in Tasmania. 1833. Dundonald *inv.* ship propelled by oscillating mercury. 1833. *Royal William*, built by Samuel Cunard at Quebec, crossed Atlantic in 22 days. 1838. Jacobi (*sic*, Prof. Jevons) *des.* electro-magnetic ship 28 ft. × 7 ft. which carried 10 persons at 4 m.p.h. on River Neva. 1839. John and M. V. Ruthven propelled a ship by jet on Union Canal near Edinburgh. 1844. 40 ft. long jet-propelled ship sailed on River Forth. 1848. Llewellyn propelled a ship electrically on lake near Swansea. 1849. Ruthven *pat.* jet-propelled ship. 1857. First steel ship, the *Ma Roberts*, built at Birkenhead, for Livingston's expedition (hull and boiler of steel). 1859. First armoured ship *La Gloire des.* by Depuy de Lôme. 1860. James Jones Aston, of London, *des.* and launched steam-tug *Saucy Jack* with 14 ft. diameter smooth discs instead of paddle-wheels, which made 6 knots per hour on River Thames. 1862. Gwynne *pat.* water-jet propelled ship *Water Witch*—a gun-boat. 1864. First sea-going steel ship built by Samuelson, of Hull, of Bessemer steel plates. 1865. Clyde ferry vessel equipped by Gwynne. 1866. Ruthven built 115 ft. *Nautilus* at Blackwall (speed of 8 knots per hour obtained). 1877. First oil tanker ship— *Zoroaster* built. 1924. ship with rotating vertical cylinder propulsion *des.* and built by A. Flettner. 1951 (July). Keel of first atom-powered ship *Savanna*

laid. 1955 (May). *Savanna* launched. 1962. *Savanna* made maiden voyage. (Will sail for $3\frac{1}{2}$ years without refuelling.) (*See also* Boat.)

Shock-absorbers, Motor-car 1903. Pneumatic type *inv.* by Maurice Houdaille. 1903. Lever type *inv.* by Truffault. 1904. "J.M." type *inv.* by Furmidge. 1909. Hydraulic type *inv.* by Houdaille. 1928. Friction type *inv.* by M. D'Albay. 1933. Double-acting, hydraulic type *pat.* by F. G. C. Armstrong.

Shoddy 1813. First made into felt at Batley, Yorkshire.

Shorthand (shorthand system *inv.* by Cicero's scribe, Marcus Tullius Tiro (106–43 B.C.); which was also used by Ennius.) 1588. Bright's system *inv.* 1682. Mason's system *inv.* but not *pub.* until 1751 by Thomas Gurney. John Willis *pub.* shorthand alphabet. 1767. Byrom *inv.* system. 1780. Mavor *inv.* system. 1786. Taylor *inv.* system. 1837. Sir Isaac Pitman *inv.* system. 1888. Gregg system *inv.*

Shot-casting *c.* 1820. Bristol plumber Watts *inv.* gravity lead shot casting, using tower of St. Mary Redcliffe church, Bristol.

Shrapnel (spherical shot) *c.* 1750. Bernard Forest de Belidor (*c.* 1678–1761) *inv.* "globes of compression." 1784. English Maj. Henry Shrapnel (1761–1842) *inv.* first known (*sic*) "spherical shot."

Shunting-yard, Railway 1879. Footner *inv.* first gravity shunting-yard at Edge Hill, Liverpool.

Shuttle, Flying 1733. *Inv.* John Kay (1704–64).

Sickle *c.* 6000 B.C. Sickle of this

date made of bone set with flints found during Palestine excavations.

SIEMENS, Ernest Werner von (Ger) (1816–92) 1866. *Intro.* so-called dynamo principle in which the magnetic field which generates electrical current is produced by the current itself.

Sifting Machines (*See* Bolting machines.)

Signalling, Railway 1834. Crossbar and lamp signals first used. 1841. Semaphore arm *inv.* by Chappé (Fr) in 1793 adopted for railways. 1842. First signal telegraph line used between Paddington and Slough (Paddington to West Drayton, 1839). 1844. Time-interval "block" system *inv.* by Sir W. F. Cooke and installed on the Norfolk Railway, of Great Britain. 1844. Wire point and signal interlock *inv.* C. F. Whitworth. 1846. First (*sic*) point and signal interlocking at Bricklayer's Arms Junction, London. 1854. "Block" signalling system *intro.* 1860. A. Chambers, of Bow, London, *inv.* "stops" or "checks" connected with signal and point levers. 1860. M. A. F. Menons, of Paris *inv.* heat compensation device for signal wires Chambers' system of interlocking *intro.* on North London Railway. 1867. John Saxby, of Brighton and James Easterbrook jointly *inv.* spring catch system of interlocking. 1875. William Sykes *inv.* "lock-and-block" system. 1893. First automatic signals (Sykes) *intro.* on Liverpool Overhead Railway.

Silhouette 1759. Originated by Étienne de Silhouette, French Minister of Finance.

Silicon (element) 1810. *Disc.* by J. J. Berzelius (1779–1848). 1823. Prepared pure by Berzelius.

Silicone (resins) 1941. Silicone resins made first appearance in U.S., after having been pioneered in England by F. Kipping.

Silk (throwing and winding machines) 1717. Silk throwing machine *inv.* (*sic*) by John Lombe, who actually copied design from the Italians and erected a silk mill at Derby; the first "factory" in England, which was *des.* by M. Crochet (Fr). 1775. Silk winding machine *inv.* by Lyons textile designer Phillipe de Lasselle, of Sessel. 1776. Silk winding machine *inv.* by civil engineer Leture, of Lille. 1795. Silk winding machine *inv.* by Biard, of Rouen.

Silk, Artificial 1664. Dr. Robert Hooke (1635–1703) put forward ideas for production of artificial silk. 1734. Réne de Réaumur (Fr) (1683–1757) suggested *prod.* of artificial silk from gums and resins. 1855. Georges Audemars (Swit) took out first *pat.* for *prod.* of artificial silk. 1884. Count Hilaire de Chardonnet (Fr) (1839–1924) *pat. prod.* of artificial silk. 1892. Viscose process *pat.* by Edward Bevan and Charles Cross. 1899. Spinnerets for 1892 process *pat.*

Silver *c.* 4000 B.C. *Disc.* 1300. First silver groschen struck at Kuttenberg. 1831. Pattison *inv.* process of de-silverization of lead. 1851. Pattison process replaced by that of Parkes.

Silver Peroxide and Sub-

oxide *Disc.* by Freidrich Wohler (Ger) (1800–82).

Siphon *c.* 1450 B.C. Shown in use on Ancient Egyptian wall paintings. 2nd cent. B.C. Mentioned by Heron, of Alexandria in his *Spiritalia*.

Siren 1819. *Inv.* by Cagniard de la Tour (Fr) and later *imp.* by Félix Savart (Fr).

Skeleton, Human Fossil 1823. First one found in the loess of the Rhone valley and shown to scientists by Aimé Boué. 1856. Neanderthal skull found near Düsseldorf. 1865. Human jawbone found at La Nuette, Belgium. 1887. Two human skeletons found at Spy, near Namur, Belgium. 1891. Neanderthal human skeleton found at Trinil, Java. *c.* 1895. Fragments of 40 human skeletons found in cave at Chou-kou-tien, near Pekin by Davidson Black and Pei (Ch). 1912. Skull found at Piltdown, Sussex and later proved to be a fake. 1933. Human skeleton found at Steinheim au de Murr, near Stuttgart. 1935. Skull found by A. T. Marston in gravel-pit near Gravesend, England. (The "Swanscombe Skull.") 1947. Two human skeletons found at Fontechevade, Fr. by Mlle. Henri-Martin. 1948. Complete lower Miocene skull found on Rusinga Island, Lake Victoria by Mrs. Leakey.

Skis Rock drawing of late Neolithic period found in Northern Norway showing man on skis.

Skyscraper 1890. First one built in Chicago. 1896. First steel-frame building (a warehouse) built in England at West Hartlepool.

Sledge Used in Scandinavia in the Mesolithic Age and in Ancient Egypt and Mesopotamia to transport stones and timber.

Slotting Machine (cotter-drill) 1856. *Inv.* by W. P. Batho, of Bordesley Ironworks, Birmingham. 1861. First used by Holtzapffel. 1862. *Imp.* by James Nasmyth.

Smallpox 569. First heard of at the Siege of Mecca. 900. First heard of in England.

SMEATON, John (1724–92) 1765. *Inv.* water-wheel-driven cylinder-boring machine and experimented on models to determine the best form of water-wheel to give the most energy as a prime-mover.

Smelting, Metal *c.* 2300 B.C. Tuycre smelting furnaces of this date found at Telloh, which suggest the use of bellows. (*See also* Iron.)

Smoke-jack 1551. *Inv.* by Jerome Cardanus (Cardan) (1501–1575).

Smoke-prevention 1715. First known example of variable smoke-prevention damper used in France—a pivoted cowl mounted on a chimney.

Soap Late 4th cent. Pliny mentioned that he preferred German to Gallic soap (made of crude natural soda and fats). 1162–1231. *Desc.* by Abdal Latif (Arab). 1524. First made in London (earlier in Bristol). 1711. Duty on soap imposed in England. 1853. Duty abolished. 1907. Frank Giffin Carter *inv.* an inlaid soap with inset abrasive of marble-flour.

Soda 1787. Nicholas Leblanc, physician to the Duke of

Orleans *inv.* process for making sulphate of soda from salt. (Committed suicide in 1806.) 1811. Ammonia-soda process evolved by A. J. Fresnel (Fr). 1825. Leblanc process *intro.* into England by Charles Tennant. 1838. J. Hemming and H. G. Dyer jointly *pat.* process similar to that of E. and A. Solvay (1861). 1854. Schlössing and Rolland set up works in Paris to operate the Hemming-Dyer process. 1861. Ernest and Alfred Solval (Bel) file ammonia-soda process *pats.* 1863. Couillet builds works to operate Solvay process. 1872. Solvay rights acquired by Ludwig Mond for G.B. 1880. Solvay process started in Germany. 1886. Solvay process started in U.S.

Soda-water 1820. Chemist Charles Cameron, of Glasgow, *inv.* apparatus for producing soda-water. (*See also* Mineral waters.)

Sodium (element) 1736. Sodium proved to be distinct from potassium by Duhamel (Fr). 1807. *Disc.* by Sir Humphry Davy (1778–1829). 1885. Manufacturing process *inv.* by Hamilton Young Castner (U.S.). 1894. New process *inv.* jointly by Carl Kellner (Aus) and Cartner U.S.).

Sodium Hyposulphite ("Hypo") 1799. *Disc.* by Chaussier (Fr).

Solar Power 1875. Solar power plant with boiler *inv.* Mouchot (Fr). 1912. Solar power plant *des.* C. V. Boys.

Solar System 500 B.C. Theory of solar system *prop.* by Pythagorus. (Other theories *prop.* by Ptolemy Claudius (2nd cent.

A.D.) and Nicholaus Copernicus (1473–1543).

Soldering 2500 B.C. Hard soldering used in Ur of the Chaldees to join sheet gold. 1838. Soldering (or "burning") of lead with hydrogen *desc.* by Desbassaynes de Richemond (Fr).

Sonnet 1024. Musical form *inv.* by Guido d'Arezzio.

Sound (Sand) Figures 1787. *Inv.* by Chladni, who raised the study of sound to a scientific level.

Sounding Machine 1773. Capt. C. J. Phillips gave first *desc.* of a deep-sea sounder when returning from Spitzbergen. 1875. Lord Kelvin (William Thomson) (1824–1907) *inv.* silver chromate sounder.

Sound, Stereophonic Reproduction of 1881. Clément Ader (Fr) *pat.* in Germany an "improved telephone equipment for theatres" which was used in Paris the same year. 1925. System *imp.* by Kapeller.

Sound, Speed of *c.* 1700. Estimated by Sir Isaac Newton at 968 ft. per second; and later by Römer (1,172 f.p.s.); Cassini (1,172 f.p.s.); Gassendi (1,473 f.p.s.); and J.C.F. Sturm (1,435 f.p.s.).

Spark Ball-discharger (radio) *Dev.* by Prof. Righi, of Bologna.

Speaking-trumpet 335 B.C. Used by Alexander the Great. 1652. One *const.* by Saland from Athanasius Kircher's description. 1671. Speaking-trumpet philosophically explained by Sir Samuel Morland (1625–95). *See also* Megaphone.

Speaking-tube (telekouphonon) *c.* 1850. *Inv.* Francis Wishaw, of Adelphi, London.

Specific Gravity Archimedes (*c.* 287–212 B.C.) made first attempt to measure specific gravity. 1603. Specific gravity of metals first determined by Ghetaldi (Arab) (1566–1627).

Spectacles 1268. Mentioned by Roger Bacon (*c.* 1210–*c.* 1293). 1270. Mentioned by Marco Polo as being used at Kublai Khan's court. 1285. *Inv.* (*sic*) by Italian monk. 1352. First picture to depict spectacles painted by Thomasso da Modena. 1480. St. Jerome (patron saint of the Spectacle-makers' Guild) painted using an eyeglass by Dominico Ghirlandajo. 1517. First picture depicting Pope Leo X using a concave lens for myopia painted by Raphael. 1608. Johannes Lippershay (Hol) (*d.* 1619) credited with *inv.* of spectacles. 1784. Bifocal spectacles *inv.* by Benjamin Franklin (1706–90). 1884. Cemented bifocal lenses *iuv.* 1908 Fused bifocal lenses *inv.* 1910. One-piece bifocal lenses *inv.* 1840. Rimless spectacles *inv.*

Spectro-heliograph 1890. *Inv.* H. A. Deslandres (Fr).

Spectroscope Sir Isaac Newton (1642–1727) *disc.* that refraction of light is always accompanied by dispersion into light of various colours. 1752. Thomas Melville (Scot) first *obs.* spectral lines. 1802. W. H. Wollaston (1766–1828) *disc.* dark lines on solar spectrum. 1814–17. Joseph von Fraunhofer (Ger) made spectroscope and assigned letters to element-indicating lines of the spectrum.

1861. *Imp.* spectroscope *des.* by R. W. Bunsen and G. R. Kirchoff. 1865. Lecocq Boisbaudran erroneously explained spectroscopic phenomena. 1867. Anders Jonas Ångström (Swed) devised the unit of spectral wavelength named after him. 1878. J. C. Maxwell published his theory of spectroscopy. (Sir David Brewster (1781–1868) *inv.* the lenticular spectroscope.) (*See also* Spectrum analysis.)

Spectroscope, X-ray *Inv.* by Sir William Bragg (1862–1942) and his son.

Spectrum Aanalysis 1800. Infra-red spectrum *disc.* by Sir William Herschel. 1801. Ultra-violet spectrum *disc.* by J. W. Ritter (1776–1810). 1859. *Disc.* by R. W. Bunsen and G. R. Kirchoff.

Speedometer, Motor-car 1904. Governor type *inv.* L. E. Cowey. 1909. "Watford" type *inv.* A. E. Rutherford and R. B. North. 1910. Magnetic type *inv.* J. K. Stewart.

Spermatoza (biology) 1679. First *obs.* and depicted by Anthony van Leeuwenhoek (Hol) (1632–1723).

Sphygmomanometer Use *intro.* by Pierre Carl Potain (1825–1901), of Paris. *Intro.* into England by Dr. Clifford Albutti (1836–1926), of Cambridge. (Recording sphygmo-manometer, or sphygmograph *inv.* by Dudgeon.)

Spinet 1440. Square spinet first mentioned.

Spinneret (viscose) 1842. *Inv.* Louis Schwabe, an English silk-weaver.

Spinning, Ring 1828. *Inv.* by

John Thorp, of Providence, Rhode Island. Post–1850. First decisive impact of ring spinning on textile industry.

Spindle "Flyer," (spinning-wheel) 1530. *Inv.* by Johann Jurger, of Wattenbuttel, Brunswick.

Spinthariscope *Inv.* by Sir William Crookes (1832–1919).

Spirit-level 1666. *Inv.* J. M. Thevenot (Fr) (–1692).

Splint, "Thomas" (surgery) *Inv.* Hugh Owen Thomas (1834–91), of Liverpool.

Splint, "Long" *Inv.* Dr. Robert Liston (1794–1847(.

SPODE, Josiah 1880. Perfected process of making English bone-china (*q.v.*).

Spoon 1490 B.C. Earliest mention of spoons—Numbers vii. 26.

SPRAGUE, Frank Julian 1884. Responsible for early *dev.* of the electric motor, particularly in respect of the street tramway car.

SPRENGEL, Hermann (Ger) 1865. *Inv.* the rotary, mercury vacuum-pump.

Springs (mechanics) Philo of Byzantium (*c.* 250 B.C.) suggested use of bronze springs to power catapaults. *c.* 1665. Road-coach springs *intro.* into England. 1805. Semi-elliptical, laminated springs *pat.* by Obadiah Elliot. 1816. Atmospheric (steam) springs *pat.* by George Stephenson and Losh.

Square, Set 1100 B.C. In use by the masons of Ancient Egypt.

Staff, Back (surveying instrument) 1607. *Inv.* by John Davis, of Sandridge, Devon.

Staining, Microscopic 1885. Possibility of staining living tissues with methylene blue *disc.* by Ehrlich (Ger).

Stamp, Postage 1840. *Inv.* by Sir Rowland Hill.

Star, Double First one *disc.* by Sir William Herschel (1738–1822).

Star, Variable 1596. First one ("Mira"—omicron Ceti) *disc.* by Fabricius Aquapendente.

STARLEY, James K. 1885. *Dev.* the direct-steering, free-wheel, braked safety bicycle— the "Rover."

Stearine *c.* 1823. *Disc.* by Michael Eugene Chevreul (Fr) (1786–1889), thus founding the candle industry. 1825. J. L. Gay-Lussac (1788–1850) founded a stearine factory which failed in Paris. 1832. Chevreul's *disc.* successfully applied by de Milly, who commenced to make candles near Barrier de l'Etoile, Paris. Process later *imp.* by Wilson and Tighlmann. (*See also* Candle.)

Steel 1740. Doncaster clockmaker Benjamin Huntsman *inv.* process of making copper-steel and nickel-steel by chemically controlling the mixture and re-heating it in a reverberatory furnace. 1750. T. O. Bergmann (Swed) realized the importance of the carbon content on the hardness of steel. 1851. William Kelly (U.S.) built his first converter; his claims as to the method used were, after his bankruptcy, made over to Sir Henry Bessemer. 1860. Sir Henry Bessemer (1813–98) *inv.* his tilting converter. 1865. Robert Mushett *inv.* self-hardening tool-steel. 1875. S. G. Thomas and Percy Gilchrist eliminate phosphorus from steel by *intro.* limestone fire-bricks. 1876. Bessemer *inv.* process for making cast-steel. 1882. Sir Robert Hadfield *inv.* manganese steel. 1884. Tungsten-chrome steel *prod.* in U.S. by Elwood Haynes. (Michael Faraday *prod.* chrome-steel.) 1889. Nickel-steel *disc.* by James Riley, of Glasgow. 1904. Léon Guillet experiments with iron and chromium to *prod.* stainless steel. 1911. P. Monnartz *pat.* stainless steel. 1912–14. Austenitic steel *inv.* by Edward Mauser and Benno Strauss (Ger) and *prod.* at Krupp's Works. *c.* 1914 Henry Brearley and Elwood Haynes, of Kokomo, Indiana, U.S. *prod.* stainless steel. (Tungsten-chrome ("high-speed") steel *inv.* by F. W. Taylor (U.S.) and Maunsel White (U.S.).) Henry Le Chatelier (1850–1936) *inv.* reverberatory gas furnace with heated air intake to make steel.

Steel, Continuous Hot-strip Rolling of 1892. A steel-mill at Teplitz, Bohemia was using this method. 1902. Charles W. Bray *des.* a hot-strip rolling-mill. 1923. John B. Tytus *des.* an *imp.* mill.

Steelyard (*See* Balance.)

Steering-gear, Motor-car 1714. Compensated gear *inv.* by Du Quet (Fr). 1772. Four-wheel steering wagon *inv.* by William Bailey. 1818. Du Quet's gear re-*inv.* by George Lenkensperger, of Munich and *pat.* in England for use on horse-drawn vehicles by Rudolph Ackermann, whose name the modern steering-gear bears. 1828. Stub-axles *inv.* by Nathan Gough, of Salford, Lancashire. 1838. Four-wheel steering re-*inv.* by Joseph Gibbs and A. Chaplin. 1873. Léon Bollée (Fr) *inv.* two-pivot cam steering-gear for his steam road coach "L'Obeissante." 1878. Jeantaud (Fr) modified the "Ackermann" gear to produce the Ackermann-Jeantaud gear. 1914. H. Marles *inv.* his steering-gear. 1921. R. Bishop *inv.* his gear. 1931. "Burman" steering-gear *inv.* by J. G. Douglas.

Steering-gear, Ship's 1866. Hydraulic gear *inv.* by Clarke and Esplen. 1867. J. McFarlane Grey *des.* steam power steering for the "Great Eastern." 1877. Beaumont *pat.* steam steering-gear.

Stencil Procopius (*d.* 560), in his *Historiae Arcana*, records that the Emperor Justinius, not being able to write, had the letters "JUST" cut in holes in a thin board to lay upon paper to direct his pen.

STEPHENSON, George (1781–1848) *c.* 1825. Pioneered the *dev.* of the steam railway locomotive, and *const.* many early English railways.

STEVIN (STEVINIUS), Simon (Hol) (1548–1620) *c.* 1585. Proved the law of the inclined plane and tested, with Jan Cornets de Groot at Delft, the truth of Aristotle's proposition that the time of fall of a body through a given distance was inversely proportional to its weight.

Stereochromy (water-glass painting) *Inv.* by von Füchs (*d.* 1856).

Stereochemistry 1848. Originated by Louis Pasteur (Fr)

(1822–95) during his work on tartaric acids.

Stereometer (liquid measurer) 1350. *Inv.*

Stereographic Projection *Inv.* by Ptolemy (100–161).

Stereoscope 1838. *Inv.* by Sir Charles Wheatstone (1802–75). 1860. Compound stereoscope, with endless bands of pictures *inv.* by S. Czugajewicz, of Paris.

Stereotyping Late 17th cent. In use in Paris. 1725. Used by William Gedd, a Scots goldsmith and his son. (Method with plaster of paris.) 1728. Papiermâché, or wet mat process *inv.* by Claude Genoux (Fr). (Flong process.) Firmin Didot (Fr) (1764–1836) *imp.* earlier processes. 1850. Gerard John de Witt *pat.* method of casting curved stereo plates on cylinders.

Stethoscope 1761 (?1816). Wooden tube stethoscope *inv.* and first used for ausculation by René Théophile Hyacinthe Laënnec (1781–1826), of Quimper, Brittany. Binaural stethoscope *inv.* and pioneered by Dr. Austin Flint (1812–1866), of New York.

Still 1850. Maxwell Miller, of Glasgow, *inv. imp.* still for distilling and rectifying spirits. *See also* Distilling.

Stirrups 853 B.C. Assyrian bronze doors of this date show flat, board stirrups in use. A.D. 100. Big-toe stirrups *inv.* in India. 477. Foot stirrup first mentioned in China.

Stocking Knitting Machine 1589. Stocking-frame *inv.* by William Lee, of Nottinghamshire. 1590. Spring-beard needle *inv.* 1764. Stocking knitting machine *inv.* by Nottingham weaver. 1847. Latch-needle *inv.* by Matthew Towsend, of Leicester. 1860. W. E. Newton *inv.* machine to shape heel and toe in a continous operation. 1864. First loom to fully fashion stockings *inv.* by Cotton.

Stokers, Mechanical 1816. John Gregson *inv.* mechanical coal thrower. 1816. John I. Hawkins and Emerson Dowison *pat.* underfeed stoker. 1819. William Brunton *inv.* revolving grate. 1822. Manchester blacksmith John Stanley *inv.* sprinkler stoker. 1833. Richard Holm *pat.* first practical underfeed stoker. 1834. J. G. Bodmer (Swit) *pat.* horizontal, perforated cylinder stoker. 1841. John Jukes *pat.* chain stoker. 1845. J. G. Bodmer *pat.* "drunken screw" stoker. 1860. Jukes *pat. imp.* mechanical stoker. 1874. John West, of Maidstone, Kent, *pat.* mechanical stoker. 1882. Babcock and Wilcox installed first mechanical stoker, watertube boiler at Hamilton Palace Colliery. (*See also* Boilers.)

STOKES, Sir George Gabriel (1819–1903) *Intro.* the word "fluorescence" in his theoretical surveys of the field of luminescence of fluorspar and quinine sulphate.

Stokes-Adams Syndrome (medical) 1827. *Disc.* by Adams and re-*disc.* by Stokes in 1842.

STÖLZEL, Heinrich (Ger) 1815. *Inv.* the musical instrument now known as the piston-horn.

Stone, Artificial 1850. Joseph Gibbs *inv.* first method of making. (*See also* Concrete and Cement.)

Stone-crushing Machine
1858. First used to prepare
material for the roads in Central
Park, New York. (*See also*
Steam-roller.)

STONEY, George Johnson
(1826–1911) *Intro*. the term
"electron."

Stovaine (drug) 1909. *Disc*.

Stratosphere 1899–1900. First
announced as being elevated at
10 km. over Europe.

Streptomycin (drug) 1944.
Disc. by Dr. Selmer A. Waks-
man (U.S.). 1946. First used in
U.S. against tuberculosis.

Streptothricin (drug) 1942.
Disc. by Dr. Selmer A. Waksman (U.S.).

Stress, Metal 1847. First
accurately calculated in framed
structures.

Striæ (In vacuo) 1843. First *obs*.

String Found in excavations
at prehistoric site at Silbury
Hill, Wiltshire.

STRINGFELLOW, John
(1799–1883) Lace-maker, of
Chard, Somerset. Partner of
William Samuel Henson (1805–
1888), with whom he collaborated in designing the motive
power for early heavier-than-
air flying-machines.

Strontium (element) 1808.
Disc. by Sir Humphry Davy.

Strychnia 1818. *Disc*. by
Pelletier and Caventon.

Strychnine 1951. Synthesized
by R. B. Woodward, of Boston,
U.S.

STUART-ACKROYD, H.
(1864–1927) 1890. With C. R.
Binney *pat*. first diesel-engine
design. (*See* Diesel engine.)

STURGEON, William (1783–
1850) 1825. *Des*. and *prod*. the
first electro-magnetic machine.

STURM, Jacob Carl Franz
(1803–55) Early worker on
physical acoustics, who in 1827
determined the velocity of
sound at 1,435 ft. per second
while working in collaboration
with Jean-Daniel Colladon.

Subconscious Mind 1889.
Disc. and investigated by Sig-
mund Freud (1856–1939).

Submarine *c*. 1495. One *des*.
by Leonardo da Vinci (1452–
1519). 1598. Cornelius Drebbel
(Hol) (1572–1634) *des*. and
const. a submarine propelled by
rowers in a diving-bell. It was
demonstrated before King
James I on River Thames in
1620. 1775. David Bushnell *des*.
and builds submarine *Turtle*.
1879. John Holland (Irish/
U.S.) *inv*. submarine completed
in 1881. Operated by one man,
it was tried on the Passiac River,
but sunk. It was, however,
raised in 1926 and is now in a
park at Paterson, New Jersey.
Holland's next effort—*Holland*,
was a success and six were
ordered by the U.S. Navy.
1885. First British submarine
launched at Barrow. 1887. Submarine tried at Southampton,
England. 1888. Lieut. Isaac
Perral remained submerged
in a submarine for one hour
in Cadiz harbour. 1892. Submarine *Sapolio* crossed Atlantic
Ocean in 68 days, arriving at
Hudva.

Submarine Cable *See* Cable,
submarine.

Sugar 325 B.C. First mention of
sugar in Western India by
Nearchus. A.D. 286. First mention of sugar in China. 1558.
Jean de Léry, calvanist minister
at Fort Coligny, Rio de Janiero,

mentions sugar. Sugar *disc.* in urine by Dr. Thomas Wallis (1621–75), of Oxford. 1900. Sugar, fructose, galactose, sorbose and mannose synthesized by Fischer.

Sugar-beet 1747. A. S. Margraaf *disc.* that beet was rich in sugar. 1801. First sugar-beet factory in world built by F. Achard in Silesia.

Sugar-boiling 1657. Five-pan plant first *desc.* by de Rochfort, at St. Christophe, Barbados. 1785. Double-bottomed sugar-boiling pan *inv.* by Thomas Wood. 1813. Vacuum-pan *inv.* by Charles Edward Howard and first used at Jagerzeile, Vienna, by Vincent Mack. 1816. Tubular steam-heated pan *inv.* Phillip Taylor. *c.* 1817. Onésiphore Pecquer (Fr) *inv.* scum-eliminator. 1817. Steam-heated oil in double-bottomed pans *inv.* by Wilson. 1824. Vacuum pans first used in France at Marseilles. 1827. Six Vacuum-pans in use in London. 1831. In the U.S. 1835. At Magdeburg, Germany.

Sulphanilamide (drug) (para-amido-benzene-sulphonamide) 1906. *Disc.* by Gelmo, of Vienna. 1930. Dr. Gerhard Domagk analysed Gelmo's 1906 production and *isol.* "Prontosil" therefrom. 1935. Domagk used Prontosil on his daughter. 1937. Re-*disc.*

Sulphapyradine (drug) (M. & B. 693) *Disc.* in the laboratories of Messrs. May and Baker, London.

Sulphur 400 B.C. Mentioned by Homer.

Sulphur Dioxide Gas 2000 B.C. Used by Ancient Egyptians for disinfecting and bleaching linen.

Sulphuric Acid 13th cent. *Prod.* by dry distillation of alum or by burning sulphur over water. First *prod.* by burning sulphur with saltpetre by quack-doctor Joshua Ward. 1746. John Roebuck *inv.* lead-chamber process. 1749. First sulphuric acid plant on Roebuck's *pat.* built at Preston Pans, near Edinburgh. 1766. First lead-chambered process factory in France. 1772. First Roebuck plant at Battersea, London. 1776. First Roebuck plant in France at Rouen. 1793. First Roebuck plant in U.S. 1831. Peregrine Phillips, a Bristol vinegar manufacturer *pat.* process for sulphuric acid *prod.* with a platinum catalyst.

SULZER, Johann Georg (1720–79) (Swit) 1760. First to mention sensation of taste experienced when silver and lead were allowed to come into contact in the mouth; thus anticipating Galvani's and Volta's discoveries.

Sundial 1450 B.C. Shadow-clock began to be used in Ancient Egypt. 713 B.C. Mentioned in Isiah xxxviii. 8. 550 B.C. *Inv.* (*sic*) by Anaximenes. 540 B.C. First recorded "hemisphere" of Berosus. A.D. 613. Sundials first set up in churches.

Sunspots First *disc.* and mentioned by Averroes (1126–1198) (Arab). 1610. Johannes Fabricus and Christopher Scheiner (1611) first to *obs.* sunspots through telescope. 1722. Influence on terrestrial magnetism first *disc.* by Graham. 1851. Relationship of sunspots with earth's magnetic field *disc.* by J. von Lamont (Ger). 1851. H. S. Schwabe *disc.* periodicy of sunspots.

Sun, Declination Tables of the 1292–95. Compiled by Robert Anglès de Montpelier (Fr) (first compilation).

Superchargers and Blowers 1866. Rotary blower *intro.* by J. D. Roots (U.S.). 1891. Gasblower or supercharger *intro.* by Dugald Clerk. 1926. René Cozette (Fr) *intro. imp.* supercharger. Supercharger first *intro.* for marine engines by Dr. Buchi (Swit).

Superheater, Locomotive Engine (*See* Locomotive, steam.)

Superphosphate (*See* Fertilizers.)

Superheating, Locomotive Engine 1845. Tried on the Great Western Railway of England.

Surface-plate 1844. *Inv.* and *prod.* by Sir Joseph Whitworth.

Surgery Paré (1517–90), father of modern surgery, *inv.* the technique of binding the arteries after amputation instead of cauterizing with a redhot iron.

Survey, Triangulated 1533. *Intro.* by Gemma Frisius (Hol). 1744–83. C. F. Cassini took 39 years to map France by triangulated survey to scale of 1 : 86,400.

Suspension System, Motorcar 1905 and 1910. P. Haincque de St. Senoch (Fr) *inv.* hydraulic and compressed-air suspension systems in which compensation was made between the four wheels of the vehicle. (*See also* Springs.)

SWAN, Sir Joseph (1828–1914) 1860. Made early experiments in production of incandescent electric lamps, first using U-shaped strips of carbonized paper.

Swash-plate Principle (mechanics) Applied in De Lavaud transmission in certain French motor-cars.

SWAMMERDAM, Jan (Hol) (1637–80) Microscopist and naturalist who, in 1669 *disc.* the modes of metamorphosis used by insects.

Switch, Oil-quenched electric 1897. *Inv.* by C. E. L. Brown.

SYKES, William 1875. *Inv.* the "lock-and-block" railway signalling system.

SYMINGTON, William (1764–1831) 1786. *Pat.* a steam road carriage with rack-and-pinion drive; also did much pioneering work on early steam-propelled boats.

Synthesis 1877. J. M. Crafts and Charles Friedel devised methods for large-scale synthesis. (*See* Crafts, J. M.)

Syringe, Hypodermic *c.* 1850. *Inv.* by Dr. Alfred Higginson (1808–84), after its use had been suggested by Claude Bernard (Fr).

T

Tachometer *c.* 1840. Bryan Donkin, of Bermondsey, London *inv.* tachometer for use in mills.

TALBOT, William Henry Fox (1800–77) Early 19th cent. Suggested the use of the bright lines seen in the spectroscope to detect substances. Applied the *discs*. of Nièpce and Daguerre to paper in photograph. 1841. *Inv.* the callotype process.

Tank (fighting vehicle) 1655. Mentioned by Marquis of Worcester in his *Century of Inventions*, as No. 31. 1420. Giovanni da Fontana sketched a military tank propelled by three rockets on rollers.

Tank (for testing model ships) 1872. Robert Froud built a 250 ft. long testing tank at Torquay.

Tantalum (element) 1801. Existence queried by Hatchett. 1802. *Disc.* by Ekeberg (Swed).

Tape and Wire Recorders 1898. Valdemar Poulsen (Dan) *inv.* first wire magnetic recorder which was shown at the Paris Exhibition of 1900. 1907. Poulsen and Peder Pedersen *inv.* D.C. biasing. *c.* 1920. W. L. Carlson and G. W. Carpender (U.S.) *inv.* A.C. biasing. Late 1920s. Kurt Stille (Ger) *inv.* steel tape recorder, and Dr. Pfleumer (Aus) *inv.* coated paper tape. 1939. Cellulose acetate tape used. 1940. P.V.C.

tape and high-frequency A.C. biasing applied—the latter by Dr. Walter Weber (Ger) and Dr. H. J. von Braunmuhl (Ger) Marvin Camras (U.S.) was one of the most prolific inventors in this field.

Tartaric Acid 1770. *Disc.* by C. W. Scheele (1742–86).

TARTINI, Guiseppe (It) (1692–1770) Italian violinist who *disc.* what came to be known as the "third tone" in musical acoustics—now known as the "difference tone."

Taxi-cab 1888. First in world in service outside Stuttgart railway station — a Daimler vehicle. 1896. First taxi-cab in France—a Roger-Benz, appeared in Paris.

Taximeter *Inv.* by A. Grüner (Ger).

Technetium (element) 1937 *Disc.* by C. Perrier and E. Segrè, in Italy.

Tedder, Hay 1800 Tedder *inv.* by Salmon, which was the basis of all modern types.

Teeth, Artificial 1710. Made by Guillemeau (Fr) from gum elemi, white wax, powdered coral and pearls. 1788. Porcelain used by Nicholas Dubois de Chément (Fr). Pierre Fauchard (Fr) (1690–1761) made artificial teeth with a sprung upper set.

Teleautograph 1814. Ralph Wedgwood perfected handwriting electric telegraph. Lenoir, Arlincourt, Jordery, and J. H. Robertson *imp*. Wedgwood's original instrument; E. A. Cooper in 1878 producing the teleautograph. (*See also* Telegraph, electric.)

Telegraph, Electric 1558. G. B. della Porta (It) (1535–1615) devised a compass-needle telegraph known as the "sympathetic telegraph." 1800. First electric telegraph *pat*. filed by Grout. 1809. Soemering *pat*. electric telegraph. 1825. Vallance *inv*. static electric telegraph. 1826. Harrison Dyer *inv*. a printing telegraph using chemically treated paper. 1830. M. H. Jacobi established first electric telegraph between Royal Palace and Ministerial office, using Arago's electromagnetic machine—not batteries. 1832. Jacobi *const*. a 21 mile electric telegraph between Russian Emperor's winter palace and summer palace at Tsarskoie, using glass tubing to insulate wires. (*Inv*. also ascribed to Baron Schiling, of Kannstadt.) 1834. C. F. Gauss and W. E. Weber (Ger) *const*. $1\frac{1}{2}$ mile electric telegraph. 1837. Steinheil (Ger) *inv*. electric telegraph. Five-needle electric telegraph *inv*. by Charles Wheatstone. 1840. Wheatstone *inv*. dial telegraph, which was later *imp*. by Breguet, of Paris. 1841. Type-printing electric telegraph *inv*. 1844. S. B. Morse transmits first electric telegraph message from Washington to Baltimore, U.S.: "What God has wrought." 1854. Thomas John, of Vienna, *inv*. printing telegraph using ordinary paper.

1858. Automatic transmission magnetic dial telegraph *inv*. 1864. J. Hawkins Simpson *inv*. printing telegraph as an improvement of the "Typo telegraph" of Bouelli (It).

Telegraph, Semaphore 1684. Dr. Robert Hooke (1635–1703) *inv*. semaphore telegraph for use with telescope. 1700. Guillaume Amontons (Fr) (1663–1705) *inv*. semaphore telegraph which was tried over short distances. 1780. Bossuet *inv*. pipe-and-water telegraph operating semaphores at each end of pipe three miles long. 1791. Claude Chappé (Fr) *inv*. semaphore telegraph which was used from paris to Lille. Models were made at Frankfurt and sent to England by William Playfair; being later adopted by the British Admiralty—the first semaphore telegraph in England.

Telegraph, Submarine 1842. S. B. Morse, an artist, conceived a sub-aqueous telegraph and connected Governor's Island with Castle Garden, New York. *c*. 1845. Capt. Taylor laid a submarine telegraph line from the Admiral's House, Portsmouth, across the harbour to the railway terminus at Gosport. 1849. South Eastern Railway telegraph superintendent C. V. Walker laid a line from two miles off Folkestone Harbour to land and thence to London via Merstham tunnel, Surrey. 1851. Submarine telegraph from Dover to Cap Grinez laid by Joseph Brett from tug *Goliath*. 1869. First trans-Atlantic submarine telegraph cable laid.

Telegraphy, Weather 1849. First tried by London newspaper *Daily News*.

Telegraphone (*See* Tape and wire recorders.)

Telephone 1821. Charles Wheatstone attempts to transmit human voice electrically. 1837. Charles G. Page (U.S.) conceived idea that electricity could carry sound. 1860. Johann Reiss (Ger) *inv.* and *const.* a telephone but did not follow up the idea. 1876. Alexander Graham Bell (Scot) *pat.* telephone; Elisha Gray (U.S.) *pat.* his telephone two hours later. 1876. Emil Berliner *inv.* a telephone. 1878. Carbon granule microphone *inv.* by David Edward Hughes. 1892. Automatic switchboard *intro.* 1906. Underground telephone cables *intro.* 1931. Transoceanic telephone services commenced. (*See also* Microphone.)

Telescope 1608. Hans Lippershay (Hol) *inv.* telescope and spectacle (*q.v.*). 1609. Galileo Galilei (1564—1642) *inv.* telescope. 1669. Sir Isaac Newton (1642–1727) *inv.* reflecting telescope. 1719. Jacques Eugène d'Allonville, Chevalier de Louville (Fr) *inv.* portable transit telescope. 1757. Achromatic telescope *inv.* John Dolland (1706–61). 1779. Diplantidian telescope giving two images, one reversed, *inv.* Edmé Sebastian Jeurat, of Paris.

Television, Mechanical 1884. Paul Nipkow *inv.* scanning disc bearing his name. 1926. John Logie Baird applied Nipkow disc in his first television transmitter. (1842. Alexander Bain proposed picture transmitter and receiver.) (*See also* Photo-electric cell.)

Televison, Electronic 1897. Ferdinand Braun, of Strasbourg *dev.* the cathode-ray oscilloscope. 1905. Julius Elster and Hans Geitel, of Wolfenbuttel, perfect an *imp.* photo-electric cell based on earlier *disc.* by Carl Hertz. 1907. Boris Rosing, of St. Petersburg proposed television system using a mechanical transmitter and a Braun oscilloscope as a receiver. 1908. A. A. Campbell-Swinton proposed use of cathode-ray tubes for both transmission and reception of images. 1923. Vladimir Zworykin *inv.* iconoscope television camera, having conceived the idea of a charge-storage camera tube in 1919. 1923 (May 21). First practical demonstration of television photograph transmission when U.S. Bell Laboratories transmitted picture of Faraday. 1924 (Dec. 1). Photo-radiogram despatched from London to New York by Capt. Richard H. Ranger. 1928. Cathode-ray image-dissector *inv.* Philo T. Farnsworth (U.S.). 1930 (Aug. 20). First transmission of television with "home picture" reception made from W2XCR, Jersey City and W2XCD, Passiac, to be received on screens at Hotel Ansonia, Hearst Building, and a house on Riverside Drive, New York. 1930 (Sept.). Station W2XAD, New Brunswick, New Jersey, transmitted first weather-map on television. 1931 (Jan.). First sight-sound dramatic television production ("The Maker of dreams") transmitted from station W9XAP, Chicago. 1939. Orthicon or camera T.V. cathode-ray tube *inv.* Rose and Iam.

TELFORD, Thomas (1757–1834) Pioneer bridge-builder and civil engineer.

Tellurium (element) 1798. *Isol.* by M. H. Klaproth. 1882. *Disc.* by Muller von Richenstein.

Temperature Chart (medical) First used by Carl Wunderlich (1815–77) (Ger).

Telpherage (overhead wire railway) 1882. Electrical system *inv.* by Henry Charles Fleeming Jenkins, of Dungeness, Kent, and installed at Glynde, Sussex. *See also* Ropeway, Aerial.

TENNANT, Smithson (1761–1815) *Disc.* the metals iridium and osmium, together with titanium oxide.

Tensile Testing Machine Leonardo da Vinci (1452–1519) sketched a machine in which a sand-fed hopper hung from the wire being tested, the entry of sand being automatically cut off by a spring when the wire broke. 1729. Machine *inv.* by 's Gravezande (Hol) *imp.* later by E. Marriotte, C. A. Coulomb, J. Rondelet and R. A. F. Reaumur. Steelyard type *inv.* by Petrus van Musschenbroek (1692–1761) (Hol). Musschenbroek *inv.* lever tensile testing machine. Amsler (Swit) *inv.* hydraulic oil-pressure tensile testing machine. *c.* 1925. E. G. Coker popularized photo-elastic stressanalysis. *c.* 1925. X-rays began to be used for non-destructive testing. (*See also* Elasticity.)

Terbium (element) 1843. *Disc.* by Mosander.

Terramycin (drug) 1949. *Disc.* by chemists at the Pfizer Chemical Company.

Terra Pinguis (Oily Earth) Theory 1669. *Prop.* by Johann Joachim Becher (Ger) (1635–82).

Terylene 1940. *Disc.* by J. T. Dickson and J. R. Whinfield. 1944. First yarn made from terelene.

TESLA, Nikola (1857–1943) (Croat-U.S.) Pioneered generation of high-frequency radio waves used in long-wave transmission, by *inv.* the rotary alternator and oscillator.

Tetra-ethyl Lead (anti-knock internal-combustion engine fuel) 1922. Midgeley and Boyd (U.S.) *disc.* that tendency for engines to detonate (knock) was suppressed by addition of minute quantitities of tetraethyl lead.

Textiles, Non-inflammable 1859. Versmann and Oppenheim (Ger) *disc.* that fabrics steeped in tungstate of sodium or sulphate or phosphate of ammonia, burnt without flame.

THABIT Ibn Querra (Arab) (826/7–901) Early 9th cent. *Prop.* the trepidation theory of equinoctial precession.

THALES of Miletus (*c* 640–546 B.C.) Pioneer of natural science who asserted that water was the prime element of all things. 600 B.C. *Intro.* geometry into Greece.

Thallium (element) 1861. *Disc.* with the spectroscope by Sir William Crookes (1832–1919).

Thaumotrope 1826. *Inv.* by J. A. Paris.

THÉNARD, Louis Jacques (Fr) (1777–1857) 1818. *Disc.* hydrogen peroxide.

Theodolite 1551. *Inv.* by Leonard Digges, whose idea was *pub.* by his son Thomas 20 years later.

Thermit (welding powder of aluminium and iron oxide) First used by Howard T. Barnes, of McGill University, Montreal.

Thermodynamics, Laws of First law *prop.* by Rudolf Clausius (1822–1888), who originated the science, as well as that of the kinetic theory of gases. 1824. Second law *dev.* by Sadi Carnot, who *dev.* the theory of heat on the older caloric theory. 1834. Carnot's theory amplified and graphically represented by Émile Clapeyron (Fr) (1799–1864). 1906. Third law enunciated by Walther Nernst (Ger) (1864–1941).

Thermo-luminescence 1676. *Obs.* by J. S. Elsholtz (Ger) when heating fluorspar.

Thermometer (*c.* 250 B.C.) Philo of Byzantium and Heron of Alexandria (A.D. *c.* 100) both *desc.* experiments based on the expansion of air by heat. Clinical thermometer *intro.* by Sanctorius (1561–1636). 1580. Savants of Padua and Florence tried to make thermometers. 1597. Thermoscope (thermometer) *inv.* by Galileo Galilei. 1600. Air thermometer *inv.* by Scorpi. 1605. Cassoni (Corsoni) *des.* heat indicator. 1609. Air thermometer *inv.* by Cornelius van Drebbel (1572–1634),

of Alkmar, Holland, as a sealed, twin-bulb instrument. 1610–11. Air thermometer *inv.* by Sanctorius of Padua. Drebbel's thermometer was modified by J. B. van Helmont (1577–1644) and by Jean Ray. 1655. Mercury thermometer *inv.* at Academy del Cimento, Florence. 1680. Mercury thermometer made by Edmund Halley (1656–1742). 1700. Guillaume Amontons (Fr) (1663–1705) *imp.* on Galileo's thermoscope. 1730. R. A. F. Reaumur (Fr) (1683–1757) *inv.* spirits-of-wine thermometer. 1733. Mercury thermometer *inv.* by M. de l'Isle, of St. Petersburg. 1742. Centigrade thermometer scale *intro.* by Anders Celcius (Swed). 1749. Märten Strömer changed Celcius's 100°–0° scale to 0°–100°. Daniel Gabriel Fahrenheit (Ger) (1686–1736) a thermometer-maker, *intro.* the thermometer scale bearing his name. 1794. J. Six *inv.* self-registering, maximum–minimum thermometer. 1799. Wet-and-dry-bulb thermometer anticipated by Leslie. 1860. W. Symonds, of Dunster, Somerset, *inv.* mercury-spirit thermometer in which iron indices were reset by raising the instrument into a vertical position. (Metallic thermometers were *inv.* by Breguet. (*See also* Pyrometers and Thermostats.)

Thermopile *Inv.* by James Prescott Joule (1818–89).

Thermoscope (*See* Thermometer.)

Thermostat 1609. Cornelius Drebbel (Hol) *inv.* thermostat for an incubator. 1830. Andrew Ure (Scot) *inv.* a bi-metal thermostat. 1864. Thermostat *inv.* by Edward Rolland. 1880. C. E. Hearson *inv.* liquid thermostat similar to Drebbel's. 1914. Bellows thermostat used by the U.S. Kelvinator Company; metal bellows having been used by L. Vidie (Fr) in his aneroid barometer of 1844. 1928. R. S. Portham *inv.* adjustable thermostat for motor-cycle engines. 1930. Thermostat first used to contol heat of domestic gas-fires.

Thimble Found in Herculaneum (A.D. 70). 1684. First made in Europe by Nicholas van Benschoten (Hol). 1695. First made in England.

THOLDE, Johann (Basil Valentine) 1546. Made first known mention of metal bismuth.

THOMPSON, Benjamin (Count Rumford) (U.S.) (1752–1814) Pioneered calorimetric investigations and *inv.* a simple photometer.

THOMSON, Sir Joseph John (1856–1940) English physicist who in 1897 *prod.* experimental proof of the existence of particles smaller than the atom.

THOMSON, William (Lord Kelvin) (1824–1907) With Rudolf Clausius (Ger) (1822–88), founded the science of thermodynamics. With d' Arsonval, *intro.* the coil galvanometer with mirror readings.

Thoraic Duct (anatomy) 1647. First *obs.* by Jean Pecquet (Fr) (1622–74), of Montpelier.

Thorium (element) 1828. *Disc.* by J. J. Berzelius (1779–1848).

Threshing Machine 1636. Sir John van Berg *inv.* threshing machine. 1722. Horse-driven

threshing machine *inv.* by mathematician Due Quet (Fr). 1731. Michael Menzies *inv.* threshing machine with rotary flails, driven by water-wheel. 1737. Horse-driven threshing machine *inv.* by Meiffen. 1743. Menzies *imp.* his machine. 1762. De Malassagny (Fr) *inv.* threshing machine. 1768. Andrew Meikle and Robert Mackell *des.* hand threshing machine. 1774. Iderton, of Alnwick, Oxley, of Flodden, and Smart, of Warwick, *des.* threshing machine. 1775. George Rawlinson *inv.* threshing machine. 1788. Andrew Meikle *imp.* van Berg's threshing machine by fitting seives and fans. 1798. Meikle *des.* steam-engine-driven threshing machine. 1800. Thomas Wigful, of Norfolk, *inv.* post threshing machine.

Thrust-pad (mechanics) A. G. M. Mitchel, of Australia, *inv.* tilting thrust-pad. (Independently *inv.* by Prof. A. Kinsbury (U.S.).)

Thulium (element) 1878. *Disc.* by P. T. Cleve (Swed).

THURSTON, R. H. (1840–1903) 1871. Pioneered the testing of lubricants, in a specially *des.* laboratory.

Thyroxin (hormone) 1914. *Isol.* by Edward Calvert Kendall.

Tidal Analyser 1872. Projected by Lord Kelvin.

Tide-mill 1044. Tide-mills operating in lagoons near Venice. *c.* 1070. Tide-mill built at entrance to port of Dover. Bernard Forrest Belidor (1697–1761) states in his book *Architecture hydraulique* that *inv.*

made by Dunkirk carpenter Perse.

TILDEN, W. A. (1842–1926) *Disc.* isoprene could be turned into artificial rubber, thereby giving the key to the industry.

Tiles 1246. First made in England.

Time-recorders 1750. Watchmen's tell-tale clocks *inv.* by Whitehurst, of Derby. 1885. Bundy (U.S.) *inv.* printing time-recorder. 1887. Benjamin Frederick Merritt (U.S.) *inv.* time-recorder. 1888. First dial time-recorder registering employees in numerical order *inv.* by Dey, of Aberdeen (? by Dr. Alexander Day (U.S.). 1888. Employees time-recorder *inv.* Willard Brundy (U.S.) (International Time-recorder Company). 1894. First time-recorder to record on employees' cards *inv.* Daniel L. Cooper, of Rochester, New York.

Time-switch 1867. *Inv.* to control gas-lamps by Dr. Thurgar. 1897. Gunning, of Bournemouth *inv.* gas-switch. 1904. Horstmann *inv.* automatic seasonally adjusting time-switch.

Tin *c.* 1750 B.C. First *disc.*

Tinder 569 B.C. *Inv.* by Anacharis.

Tin-plate 13th cent. Tinned iron used for parts of armour in Bohemia.

Titanium (element) 1789. *Disc.* in Cornwall by Rev. Gregor. 1794. Found by M. H. Klaproth in Hungary, and named by him. (W. A. Lampadius completely reduced titanium.)

Tobacco-plant 1558. Cultivated in Spain. 1586. Cultivated in England (Gloucestershire) by Virginian emigrants.

Tobacco-mill (Snuff-mill 1828. Rotary tobacco-mill *inv.* by Samuel Wellman Wright.

Tonic Sol Fa System 1812. *Inv.* by Miss Glover, of Norwich. 1828. *Inv.* (?) by M. Sudré (Fr) who gave it the name "Musical Language" or "telephony." 1844. System *dev.* by Rev. John Curwen (1817–1880). (*See also* Chevé musical system.)

Torpedo 1280–95. Syrian al-Hassan al Rammah proposed a rocket-propelled aerial torpedo. 1777. Submarine torpedo *inv.* David Busnell (U.S.). 1861–65. Electrically fired torpedo used in U.S. 1864. Robert Whitehead *inv.* self-propelled submarine torpedo. 1865. Ship "Terpsichore" sunk at Chatham by the magnetic, electrically fired torpedo tried by McKay and Beardslee. 1866. Prof. F. Abel's submarine torpedo tried at Woolwich. 1869. Whitehead *inv.* pendulum control for his torpedo. 1875. First self-propelled torpedo sunk a Turkish battleship.

Torpedo-net 1655. Mentioned by Marquis of Worcester in his book *Century of Inventions No. 11.*

Torque-convertor 1904. H. Föttinger, of Hamburg, *inv.* separate torque-convertor and hydraulic coupling. 1926. First applied to motor-cars by Reisler (Ger). 1928. First applied to motor-buses (Leyland) by Alfred Lysholm (Swed). 1933. First applied to rail-cars (Leyland).

TORRICELLI, Evangelista (It) (1608–1647) Amanuensis

to Galileo. 1643. *Inv.* the Barometer.

Torsion 1777. Theory outlined by C. A. Coulomb (Fr) (1736–1806), who in 1802 *inv.* the torsion balance. 1908. Hopkinson-Thring Torsionmeter *intro.*

Tourniquet (surgery) 1674. *Inv.* Morelli (Morel) (Fr) at Siege of Besançon. 1718. Jean Louis Petit (Fr) (1674–1750) *inv.* screw tourniquet.

Town-planning, "Gridiron" Began in Ancient Greece when Hippodamus of Miletus re-shaped the Piraeus.

Tracheotomy (surgery) 1825. *Intro.* by Pierre Bretonneau (1771–1862), of Tours, for laryngeal diptheria.

Traction-engine, Steam 1618. *Pat.* filed by David Ramsey and Thomas Wildgoose. 1842. Traction-engine *des.* by William Worby. 1845. Clayton and Shuttleworth *des.* 8 h.p. traction-engine with locomotive boiler. 1846. J. T. Osborne *pat.* John Boydell *pat.* track-laying traction-engine. 1849. Barret and Exhall, of Abingdon *pat.* Robert Willis *des.* traction-engine. 1854. John Fowler, of Woolston, Lincolnshire, exhibited steam ploughing traction-engine at Lincoln. 1856. William Bray *inv.* traction-engine with spike-fitted wheels for soft ground. 1858. Thomas Aveling builds his first traction-engine. 1864. Lotz, of Nantes, builds traction-engine and later steam omnibuses (*q.v.*). 1865. Shuttleworth *des.* first traction-engine with differential transmission. 1870. Aveling builds first all-gear-drive

traction-engine. 1878. Aveling *pat.* two-speed countershaft drive. 1880. Edward Foden builds his first two-cylindered traction-engine, and in 1887, one with two cylinders compounded. 1882. Copeland, of Philadelphia, U.S., *des.* traction-engine. 1887. Charles Burrell *intro.* traction-engine mounted on springs. 1890. Aveling *intro.* four-shaft traction-engine. *See also* Steamroller, and Road Vehicles, Steam.

Tractor, Tracklaying ("Caterpiller") 1825. Sir George Cayley (1773–1857) *inv.* tractor calling it a "universal railway." 1826. James Bryan *inv.* a tractor. 1846. John Boydell *inv. pat.* tractor. 1855–58. Burrel-Tuxford-Boydell tracklaying engine system *intro.*

Traffic Signals 1866 (Dec. 10). *Inv.* by Hodgson and set up in Westminster, near the Abbey.

Trafficators, Motor-car 1910. Mechanical trafficators *inv.* by J. H. Faulkner.

Transcendentals 1844. *Disc.* by Liouville (Fr).

Transfer Machine 1924. First one in England made for Morris Motors by Archdale.

Transfinite 1883. *Inv.* by Cantor (Ger).

Transformer, Electric (induction-coil) 1829. Joseph Henry (U.S.) (1797–1878) *disc.* principle of electric induction. 1830. Michael Faraday (1791–1867) *disc.* and *form.* theory of electrical induction, independently to Henry. 1850. Heinrich Daniel Ruhmkorff (1803–77) *inv.* the induction-coil (transformer). 1882. J. D. Gibbs *inv.* the use of transformers in series. 1887. George Westinghouse (1846–1914) *pat.* air-cooled and water-cooled transformers. 1903. R. Hadfield (1859–1940) *inv.* first transformer with silicon-steel core. 1907. M. A. Codd *inv.* the "Mira" coil igniter (transformer) for motor-cars. 1926. Delco-Remy coil (transformer) ignition system for motor-cars *intro.*

Transmutation of Elements 1919. First achieved by Sir Ernest Rutherford in collaboration with James Chadwick. (*See also* Atom.)

Traverser, Railway Track 1850. *Inv.* by Ormerod and Shepherd.

Treadle A.D. 2nd cent. In use in China. Late 12th cent. Alexander Neckham, of London, *desc.* a treadle worked loom. 1418. Treadle applied to pipe-organ as a keyboard.

TREMBLEY, Abraham (Swit) (1700–84) *c.* 1745. Biologist who made first mention of the Leyden jar, or condenser.

TREVITHICK, Richard (1771–1833) 1801. *Const.* a high-pressure steam road vehicle from Weath to Camborne Beacon, Cornwall. 1812. *Inv.* the "Cornish" boiler. 1812. *Inv.* conical ore-crushing rolls, steam winding-engines, rock-drills, steam threshing-machines and a dredger.

Trigonometry A.D. 2nd cent. First treatise written by Claudius Ptolemaeus (Ptolemy). 5th cent. Paulisa (Hindu) wrote on trigonometry. 9th cent. Ben Musa (Arab) substituted sines for chords in trigonometry. 10th cent. Abu'l Wafer (Arab) *intro.* tangent as independent function and not merely as ratio of sine to cosine. Also *dev.* trigonometry: Ibn Yunos, of Cairo (*d.* 1008); Uleg Beg (1393–1449); Purbach, of Austria (1423–61); Georg Joachim (Rhaeticus); and Bartholomaus Pitiscus (1561–1613). Hyperbolic functions of trigonometry *intro.* by Johann Heinrich Lambert (1728–77). Trignometry abbreviations, "sin," "cos" and "tan" first used by Albert Girard (Hol) (1595–1633).

Trocar (surgery)*Inv.* by Sanctorius (1561–1636).

Troposphere (*See* Stratosphere.)

Truck, Fork-lift 1830. Germ of the idea contained in S. W. Wright's *pat.* vehicle *des.* for West India Dock Company.

Trumpet (clarion) 800. *Inv.* by the Moors.

Tubbing (mine-shaft lining) 1777. *Inv.* by John Carr.

Tuberculosis (basteria) 1882. *Disc.* by Robert Koch (Ger) (1843–1910), of Woolstein.

Tubing (*See* Pipes.)

TUDOR, Frederick (U.S.) (1783–1864) 1806. Equipped brig *Favourite* to carry 130 tons of ice to St. Pierre, Martinique. *See also* Refrigeration.

Tungsten (element) 1781. *Disc.* by d'Elhujar. 1783. *Isol.* by C. W. Scheele. 1786. Brothers De Luyart *isol.* tungsten from tungstic acid. 1859. First tungsten-steel rods *prod.*

Turbine, Gas 1648. Bishop Wilkins, in his *Mathematical Magic* proposed use of gas turbine to rock cradles or turn roasting-spits. 1791. John Barber *pat.* gas turbine and compressor. 1906. Lemale and Armengaud *pat.* gas turbines. 1908. Holzwarth and Körting *pat.* gas turbines. 1909. Lake (U.S.) *inv.* pressurized fresh-air addition to improve force of jet. 1933. Leduc (Fr) *des.* jet-propelled aircraft without compressors. (*See* Aeroplanes, jet.) 1939. Heinkel flew aeroplane with turbo-jet engine. 1941. Gloster-Whittle aeroplane, with engine *des.* by Sir Frank Whittle, flew. 1941. Caproni-Campini jet aeroplane flew 300 miles in Italy. *See also* Aircraft.

Turbine, Mercury 1913. W. R. L. Emmet *dev.* two-fluid cycle heat engine using water and mercury.

Turbine, Steam 1629. Giovanni Branca pictured a steam turbine with a sufflator as boiler in a shaped human head of brass. 1641. Athanasius Kircher depicts small wind-vane turned by two steam jets. 1776. John Barber *pat.* steam-wheel. 1784. Baron von Kemperlin (Ger) *pat.* "reaction wheel," which was later brought to notice of James Watt by his partner Matthew Boulton; Watt taking out a *pat.* in England for his "steam wheel" in 1784. 1815. Richard Trevithick built his "whirling engine" of similar design to that of Heron of Alexandria (A.D. *c.* 100). 1837. Gilman *desc.* multi-cellular, radial-flow turbine in which the steam expanded in stages. 1837. Avery, of Syracuse, U.S., built steam-wheels 5 ft. in diameter; these later being *const.* by

Wilson, of Greenock, Scotland. 1838. Timothy Burstall *pat.* steam turbine with backward-curving steam jets. 1838. Heat *desc.* Heron-type steam turbine with trumpet-shaped nozzles. 1843 Pilbrow experiments with steam nozzles and calculated optimum vane-speed as 1,250 ft. per second. 1858. John and Ezra Harthan *des.* steam turbine which was the forerunner of Curtis's *des.* of 1896. 1882. Dr. Gustav de Laval (Swed) *inv.* single-nozzle steam turbine. 1884. Sir Charles Algernon Parsons (1854–1931) *pat.* his first steam turbine (10 h.p.). 1884. Robert Wilson, of Greenock, *pat.* radial-flow and axial-flow, and having both fixed and moving blades. 1885. De Laval *inv. imp.* steam turbine. 1896. Curtis (U.S.) *inv.* velocity-compounding steam turbine. 1903. Dr. Henrich Zöelly (Swit) *inv.* steam turbine for railway locomotives. Prof. Auguste Rateau (Fr) *inv.* compound impulse type steam turbine. 1910. Birger Ljungström (Swed) *inv.* radial-flow steam turbine for railway locomotives.

Turbine, Water 1743. J. T. Desaguliers (Fr) (1685–1744) put into use a reaction water turbine previously *inv.* by R. Barber. 1750–54. Reaction water turbine *inv.* by Leonhard Euler (1707–83). 1791. Reaction water turbine proposed by James Sadler. 1792. Impulse water turbine first (*sic*) proposed by John Bailey (U.S.). 1823. Outward-flow reaction water turbine *inv.* by Benoit Fourneyron (Fr). 1824. Fourneyron and Prof. Claude Burdin

(Fr) *dev.* outward-flow water turbine. 1826. J. V. Poncelet (Fr)proposed inward-flow water turbine which was *const.* in New York later. 1827. B. Fourneyron (Fr) *const.* first radial outward-flow water turbine and installed it at Pont sur l'Ognon, Saone, France. 1836. First inward-flow water turbine *inv.* by Howard and *imp.* by James B. Francis in 1847. (Also *att.* to A. M. Swain (U.S.).) 1840. Francis also *inv.* inward-flow water turbine 1841. N. J. Jonval (Fr) *intro.* parallel-flow water turbine. 1844. Boyden *inv.* diffuser for water turbines, which raised their efficiency by 6 per cent. 1852. J. J. Thomson (Lord Kelvin) *inv.* vortex-wheel. 1854. Impulse water turbine *inv.* in California. 1880. L. A. Pelton (with J. Moore) *dev.* the central-partition wheel-bucket now bearing his name. 1913. Prof. H. C. V. Kaplan (Swed) *pat.* his propeller water turbine with blade-angles adjustable by a governor. 1926. Kaplan water turbines of 26,000 h.p. in use in Sweden.

Turkish Baths 1860. *Intro.* into England.

Tychonic System (astronomy) *Prop.* by Tycho Brahe (Dan) (1564–1601).

Tympanum, Artificial *Inv.* by James Yearsley (1805–69).

Type, Printing A.D. 740. Wood and metal block printing in use in China. 1045. Movable earthenware type *inv.* by Pi Sheng (China). 1314. Movable wooden type used in China. 1392. Movable metal type in use in Korea. 1340. Movable metal type in use in China

(*inv.* ?). 1452. Peter Schöffer (Ger) cast first metal type in matrices in the West. (*Inv.* not announced until 1457.) 1465. Gothic, or black-letter type used until this date. 1476. Aldus, of Venice, casts Greek type alphabet and *intro.* italics. Johann Gensfleisch Gutenberg (1397–1468) *inv.* (?) movable type. Typefaces *des.*: Plantin, by Christopher Plantin, of Antwerp; Baskerville, by John Baskerville (1706–75); Garamond, by Claude Garamond (–1561); Bodoni, by John Baptist Bodoni (It) (1740–1813). 1826. Arabic type *inv.* by Holman Hallock (U.S.).

Typecasting Machine *c.* 1822. First machines. 1828. Thomas Aspinall *inv.* typecasting machine. 1838. First successful typecasting machine (U.S.). 1841. Typecasting machine *inv.* by Ballanche, of Lyons. 1842. Sir Henry Bessemer (1813–98) *inv.* piano-keyboard typecasting machine. 1846. Timothy Alden (U.S.) *inv.* Monotype. 1851. First successful English typecasting machine. 1859. Hattersley *inv.* typecasting machine. 1880. Dr. Mackie *des.* typecasting machine for newspaper *Warrington Guardian*. 1890. Linotype *inv.*

Typewriter 1744. Henry Mill *pat.* typewriter. 1784. Typewriter using characters for the blind *inv.* 1829. Austin Burt (U.S.) *inv.* typewriter. 1833. Xavier Progin, of Marseilles *inv.* typewriter. 1841. Alexander Bain and Thomas Wright *inv.* typewriter. 1843. Charles Thurber, of Worcester, Mass.,

U.S., *inv.* first handwriting letter-spacing typewriter. 1844. Littledale *inv.* embossing typewriter 1849. Pierre Foucault (Fr) *inv.* embossing typewriter shown in London. 1850. S. A. Hughes *inv.* blind typewriter. 1851. Sir Charles Wheatstone *des.* piano-keyboard typewriter. 1856. A. E. Beach *des.* blind typewriter. 1866. John Pratt (U.S.) *pat.* typewriter in England. 1895. J. F. Hardy *inv.* shorthand typewriter. 1910. Marc Grandjean, of Paris, *inv.* shorthand typewriter.

Typhoid Fever 1896. Sero-diagnostic test *disc.* by Georges Widal, of Paris. Bacillus of typhoid *disc.* by Georg Caffky (Ger) (1850–1918).

Tyres, Road 1845. R. W. Thompson (Scot) *inv.* first pneumatic rubber tyre. 1855. Charles Goodyear (U.S.) (1800–60) *pat.* rubber tyres (*pat.* never sealed). 1855. Uriah Scott *inv.* rubber tyres (*pat.* never sealed). 1865. Thompson *intro.* solid tyres and fitted them to his traction engines. 1881. Clincher cycletyre *inv.* by W. H. Carment. 1888. James Boyd Dunlop (1840–1921) re-*inv.* pneumatic tyres for bicycles. 1904. Tubeless tyres *inv.* by Martin. (Valve for pneumatic motor-car tyres (Schraeder) *inv.* by M. C. Schweinert and H. P. Kraft in 1914.) 1904. Carbon black in rubber outer-covers first used by S. C. Mole. 1910. First aeroplane tyres made by Dunlop. 1917. Bullet-proof tyres made by Dunlop. 1921. First use made of rayon in tyre casings. 1953. Tubeless tyres of natural rubber *intro.*

Tyrosin (chemical) *Disc.* in urine by Friederich Theodor von Frerichs (Ger) (1819–85).

Tyrothricin (drug) 1939. *Disc.* by Dr. René Dubos (Fr).

U

Ultra-microscope 1903. *Inv.* by R. A. Zsigmondy and Siedentopf.

Ultra-violet Light 1801. *Disc.* by Johann Wilhelm Ritter (1776–1810). Effect of ultra-violet light on human skin *disc.* by Neils Ryberg Finsen (Dan) (1860–1904).

Umbrella (parasol) *c.* 1578. In use in Italy. 1616. First mention of umbrella in England in comedy by Ben Jonson. 1622. In Paris. 1709. *Imp.* by Marius (Fr) similar to modern umbrella. Pre-1835. Carey *inv.* collapsible metal frame. (Later *imp.* by Deacon and J. G.

Hancock, of Birmingham.) 1850. Alapaca first used for umbrellas.

Universal Joints (flexible couplings) 1551. *Inv.* by Jerome Cardanus (It). Universal joints *inv.* by Dr. Robert Hooke (1635–1703). 1841. *Inv.* by J. G. Bodmer. 1914. Fabric "spider" universal joint *inv.* by John Hardy.

Uranium (element) 1798. *Disc.* by M. H. Klaproth (1743–1817), as oxide. 1931. F. W. Aston (1877–1946) announced detection of U.238 in the spectrograph of uranium hexafluoride. 1935. A. J. Demster (1866–) *pub.* his detection of U.235 isotope. 1939. U.234 isotope *disc.* by A. O. Nier.

Uranus (planet) 1781. *Disc.* by Sir William Herschel (1738–1822). Satellites of Uranus *disc.* as follows: Nos. 1 and 2, by Herschel in 1787; 3 and 4, by Herschel in 1790; 5 and 6, by Herschel in 1794; 7, by Lassell in 1847; and 8, by Struve in 1847.

Urea First prepared synthetically by Friederich Wöhler (Ger) (1800–82).

UREY, Dr. Harold C. (U.S.) 1932. Joint discoverer of deuterium ("heavy hydrogen").

Uric Acid 1780. *Disc.* by C. W. Scheele (Swed).

USHER, James 1849. *Pat.* the first steam-driven ploughing machine.

U-tube Manometer 1874. *Inv.* by McLeod. (*See also* Gauge, pressure.)

V

Vaccination 1780. Edward Jenner (1749–1823) conceived the idea of vaccination, which was ridiculed by the leading physiologists of the day. 1796. Jenner's first patient, a boy named Phipps, was inoculated with cowpox lymph from a pustule on milkmaid Sarah Holmes, by Jenner. 1798. Jenner's *disc. pub.*

Vacuum-cleaner 1901. *Inv.* in England by H. C. Booth.

Vacuum-flask *Inv.* by Sir James Dewar.

Valency, Chemical 1852. Assigned to atoms of the elements by Edward Frankland. 1858. August von Kekulé (Ger) (1829–96) determined valency of carbon atom as four.

VALENTINE, Basil (*See* Johann Thölde.)

Valve (tap) *c.* 1730. Reversing valve *inv.* by Jacob Leupold, of Leipsig (1674–1727) for a proposed high-pressure steam-engine with two single-acting cylinders. *Des.* used by Richard Trevithick, *c.* 1804.

Valve, Hydraulic 1713. Humphrey Potter *inv.* hydraulic valve.

Valve, Safety 1681. Steelyard type *inv.* by Denis Papin. 1718. *Imp.* by Beighton. Woolfe *inv.* loaded-plug type. 1837. Benjamin Hicks, of Bolton, Lancashire, *desc.* use of ball-valve used in his safety valve, mentioning that the type had been used in water-pumps for over 100 years. Sockl, of Lambeth, London, *inv.* diaphragm type. Darnell, of Pentonville, London, *inv.* floating type.

Valves, Steam-engine 1799. Double slide-valve *inv.* and *pat.* by William Murdock (1754–1839. 1801. Double slide-valve *inv.* Matthew Murray, of Leeds. *c.* 1820. George Stephenson and William Howe, of Chesterfield *inv.* "Stephenson Link" valve-gear. 1832. William James, of New York, *inv.* link-motion valve-gear. 1844. Locomotive valve-gear *inv.* by Egide Walschäerts (Bel). 1849. George Henry Corliss (U.S.) (1817–88) *inv.* valve-gear. 1855. Green *inv.* valve-gear.

Valve, Thermionic 1904. Two-electrode valve *inv.* by J. A. Fleming. 1907. Three-electrode valve *inv.* by Lee de Forest (U.S.).

Valve (of Veins) *c.* 1680. *Disc.* by Fabricius ab Aquapendente, William Harvey's (1578–1657) tutor.

Vanadium (element) 1801. Del Rio *disc.* vanadium in lead ore, but called it erythronium;

which was later proved by Freidrich Wöhler (1800–82) to be vanadium. 1830. *Disc.* by Sefström combined with iron ore and classed as an element. 1865. *Disc.* in Cheshire copper-bearing beds by H. E. Roscoe.

Van Allen Belts (astronomy) 1958. *Disc.* by Dr. James Van Allen (U.S.).

VARIGNON, Pierre (Fr) (1654–1722) 1725. First recognized significance of the parallelogram of forces.

VARLEY, S. Alfred (1832–1908) 1866–67. With Werner and Carl Siemens *dev.* a practical self-exciting dynamo.

Vasomotor Mechanism (anatomy) 1852. Vaso-constrictor fibres *disc.* by Claude Bernard of St. Julien, Lyons (1813–78). 1867. Effect of nitrites on production of vaso-dilation *disc.* by Sir Lander Brunton (1847–1916), of St. Bartholomew's Hospital, London.

VAUCANSON, Jacques de (Fr) (1709–82) Outstanding inventor. 1748. *Inv.* wheel-driven silk-weaving loom. Also lathes and screw-cutting machines.

VAUQUELIN, L. N. (Fr) *Disc.* beryllium, 1797; chromium, 1798; also quinic acid, asparagine and camphoric acid.

Vector Analysis Method founded by A. F. Mobius (Ger) (1790–1868) and H. G. Grassmann (Ger) (1809–77). (Simon Stennius originated vector analysis as parallelogram of forces.)

Vehicle Suspension, Pneumatic 1812. Joseph Bramah (1748–1814) *pat.* idea. *c.* 1825.

George Stephenson and W. Losh *pat.* steam suspension system for railway locomotives. 1834. W. H. Barlow *inv.* air-cushion vehicle suspension for railway coaches. 1839. Moses Poole *inv.* metal-box bellows vehicle suspension. 1842 William Henry James *inv.* cushions of caouchouc to insert between frame and axles of railway coaches. (Other early pneumatic vehicle suspension system; *inv.* by Rayner, Bell, Hancock, Macintosh, Lyall and Richardson.) 1905. P. H. de St. Senoch (Fr) *inv.* equalized pneumatic vehicle suspension system; *imp.* in 1910. 1911. I. Cowles and E. H. McDowall *inv.* hydro-pneumatic, oil equalized system.

Velocipede (*See* Cycle.)

Velvet 1272. First mentioned by Joinville. 1399. "velveto" mentioned in King Richard II's will. 1680. First manufactured in England.

Veneering 1600 B.C. Art of veneering practised in Egypt.

Ventilation 600 B.C. Ventilation by draught of fires used at Lorion silver-mines by the Athenians. 1550. Georgius Agricola (1490–1555) *desc.* and *illus.* many methods of mine ventilation, including by fans and bellows. 1740. Bellows ventilation installed on ship *Sorbay. Inv.* by Stephen Hales (1677–1761). Also used for granaries and prisons; displacing 25,000 cu. ft. per hour. 1748. French highway engineer Pommier equipped Hotel des Invalides, Paris, with an *imp.* ventilation system. 1749. Sutton *inv.* non-bellows ventilation sys-

tem. 1752. Hales applied windmill to ventilation of dwellings. 1819. Marquis de Chabanne planned ventilation system which was tried in London theatres.

Venturi Meter 1887. *Inv.* Clemens Herschel (U.S.) who named it after Italian scientist Venturi.

Venturi-tube 1787. *Inv.* (?) by Matthew Boulton (1728–1809). 1792. *Pat.* applied for by Whitehouse. 1797. *Inv.* by Venturi (It). (*See also* Hydraulic ram and Flow-meter.)

Venus (planet) 1611. Phases of Venus *disc.* by Galileo. 1639. Transit of Venus first *obs.* by telescope by Horrocks. 1667. Diurnal rotation of Venus *disc.* by Giovanni Dominico Cassini (It) (1625–1712).

VERMUYDEN, Cornelius (Hol) (*d.* 1665) *c.* 1650. Famous civil engineer and *inv.* who drained the fens north of Cambridge.

VERSALIUS, Andreas (Bel) (1514–64) 1543. Wrote treatise *De Fabrica corporis humanii*, and therefore accepted as the father of modern anatomy.

Vesta (planet) 1807. *Disc.* by Heinrich Wilhelm Matthaus Olbers (1758–1840), of Bremen.

Viaduct (*See* Bridge.)

Vibro-cardiagraph Cathode-ray type *dev.* by Kauntz (Ger).

Vice 420 B.C. *Inv.* (*Leg.*) by Archytas of Tarentum, pupil of Pythagorus. 1790. Quick-grip vice *inv.* by Joseph Bramah (1748–1814).

Views, Dissolving (lantern) *inv.* by H. L. Chide (1780–1874).

VILLAUME, Jean-Baptiste

(Fr) 1849. *Inv.* 13 ft. high octobasse (musical instrument).

VINCI, Leonardo da (It) (1452–1519) One of the greatest geniuses who ever lived. Painter, sculptor, architect, civil and mechanical engineer. *Inv.* paddle-wheel, breech-loading cannon, mincing machine, ornithopter, helicopter, parachute, and many other *invs.* Also made *discs.* in optics, perspective, friction, heat and astronomical theory. Unfortunately for technology, his *invs.* and *discs.* were not *pub.* until many hundreds of years after his death. (*See* Individual items.)

"Vinyon" (man-made fibre) 1933. First *prod.* in U.S.

Violin 1200. First mentioned. *Intro.* into England in time of Charles I. Finest instruments made by Stradivarius, of Cremona between 1700 and 1722.

Virus 1885. Louis Pasteur (Fr) (1822–95) made first virus inoculation against rabies and hydrophobia on Alsatian boy, Joseph Meister. (*See* separate headings.)

Viscosity 1842. Laws governing flow of viscous liquids *disc.* by J. L. M. Poiseuille.

Vision, Persistence of A.D. 130. Mentioned by Ptolemy in his second book of optics.

Vitamins *Disc.* by Gowland Hopkins. 1911–12. Casimir Funk obtained a crystalline substance from rice polishings and named it "vitamine."

Vitamin A 1915. *Disc.* by E. V. McCullum. 1931. *Isol.* by P. Karrer. 1937. Synthesized by Kuhn and Mirris.

Vitamin B2 (lactoflavin) Synthesized by Dr. A. H. Cook in Germany.

Vitamin B12 (cobalt complex) 1948. *Disc.* by Rickes.

Vitamin D 1882. Trousseau recognized that cod-liver oil was a cure for rickets. 1890. Pulm suggested that sunlight had an antirachitic action. 1919. Huldschinski first employed ultra-violet lamp to cure rickets. 1921. Hess and Unger demonstrate antirachitic effect of sunlight.

Vitamin E 1922. *Disc.* by Herbert McLean Evans.

Vitamin K 1939. *Isol.* from alfalfa and putrefying fishmeal by McKee and later same year synthesized by Binkley and Feiser. 1939. Anti-haemorrhagic property *dem.* by Almquist and Klose.

Vitamin P First obtained as crystals from lemon-juice and Hungarian red peppers by Gyorgyi Szent.

VITRUVIUS Pollio, Marcus (A.R.) (*c.* 50–26 B.C.) Famous practical engineer and *inv.* who contrived and explained the features of many different types of early cranes and hydraulic gear.

VIVANI, Vicenzo (It) (1622–1703) Famous early civil engineer who, in 1679, carried out Leonardo da Vinci's control plans for the River Arno, 84 years later.

VOGEL, Hermann Carl (Ger) (1841–1907) 1879. Carried out much early experimental work connected with spectroscopic analysis.

VOLTA, Count Alessandro

(It) (1745–1827) 1800. *Inv.* the electric battery, or pile bearing his name.

Voltaic Cell 1800. *Inv.* by Count Alessandro Volta.

VOSSIUS, Isaac (Hol) (1618–69) 1662. Discussed many hitherto unsolved optical problems in his book *On the Nature and Properties of Light.*

Vote-recorder 1861. First *pat.* by Thomas Alva Edison (U.S.) (1847–1931).

VULTURIUS, R. 1472. Wrote and *pub.* the first book on engineering.

WAAGE, Peter (1833–1900) 1867. With C. M. Guldberg formulated the fundamental law of chemical kinetics—the "law of mass action."

WAALS, Johannes Diederick van der (Hol) (1837–1923) 1872. Formulated a new law of gas physics which extended the older law *prop.* by Gay Lussac.

WAGNER, Ernest (Ger) (1829–89) 1862. First to employ the terms "carbohydrate" and "lipoid."

Wagon, Travelling Mill A.D. 340. In use in China.

WALKER, Sears Cook (U.S.) (1805–53) *c.* 1845. Astronomer associated with the technicalities of the disparagement of Messrs. Adams and Leverrier's theoretical work which ended in the *disc.* of the planet Neptune.

WALLIS, John (1616–1703) 1668. With Sir Christopher Wren a champion of Descartes's "collision theory" of light. Did much pioneer mathematical work leading to the ultimate *inv.* of the Calculus (*q.v.*).

Washing Machine (crockery and textile) 1850. Cylindrical type with agitated water *inv.* Joel Houghton (U.S.). 1858. First rotary type *pat.* by Hamilton E. Smith, of Pittsburg, U.S. 1863. First self-reversible type *inv.* by H. E. Smith. 1907. First self-controlled type—the "Thor" *pat.* by Alva J. Fisher (U.S.).

Wassermann Reaction (medicine) Blood test for syphilis *disc.* by von Wassermann, of Berlin (1866–1925).

Watch 1504 Oldest watch now in existence kept at Imperial Hall, Philadelphia, Pa., U.S. Made by Peter Henlein, or Hele, of Nuremburg, where the first watches were made. (This watch is the earliest example of a true spring-driven "clock-watch.") 1580. First watch made in England. 1630. First watch made to have a glass. 1655. Pocket time-piece mentioned in Marquis of Worcester's *Century of Inventions*, *No. 78*. 1658. Robert Hooke *inv. (sic)* spring pocket-watch. 1675. Christiaan Huygens *des.* a spiral spring-balance which was a success. (Robert Hooke and Abbe de Hautfeuille both claim credit for this *inv.*) 1686. Rev. Edward Barlow *inv.* repeating watch. 1688 (? 1696). Daniel Quare *inv.* repeating watch. 1704. Jewelled pivot-holes made of sapphires *inv.* by Nicholas Faccio de Duiller (Swit) and Peter Jacob Debaufe, a Swiss watchmaker. 1755. Thomas Mudge *inv.* the "English lever" watch escapement mechanism. 1766. Le Roy Pierre (Fr) *des.* a chronometer with most features of a modern watch, including a bi-metallic balance-wheel of brass and steel. 1780. Luis Recordon *inv.* self-winding watch. (A watch running for one year without rewinding also made by Geneva watchmaker Jean Romilly (1714–96). 1789. Gong repeating watch *inv.* by Recordon. 1790. First wrist-watch made by Jaquet Droz and Paul Leschot, of Geneva. Pre-1850. Keyless wind and hand-set watches *inv.* in Switzerland. 1865. Roskopf (Swit) *des.* first cheap watch on mass-production lines. 1880. Buck (U.S.) *des.* the Waterbury Watch. (Louis Breguet (Fr) (1780–1820) *inv.* the water hair-spring bearing his name.)

Water, Decomposition of (1784. Composition of water

disc. by Henry Cavendish.) 1789. Joan Rudolf Deiman (Hol) and Adriaan Paetz von Troostwijk (Hol) *disc.* decomposition of water electrolytically. 1800. William Nicholson and Anthony Carlisle use a voltaic pile to decompose water into oxygen and hydrogen. (*See also* Electrolysis.)

Water, "Aërated" 1807. *Pat.* by Henry Thompson. 1832 and 1847. *Pat.* by F. C. Bakewell. 1840. *Pat.* by Tylor. (*See also* Mineral waters, natural and artificial.)

Water, Chlorination of 1896. Process first used at Pola, Italy, to stem typhoid fever epidemic. 1897. First installed in England at Maidstone, Kent, for the same reason.

Water-closet *c.* 1460. *Desc.* by Sir John Harrington. 1778. Flush type *inv.* by Joseph Bramah. 1782. Water-seal trap for, *inv.* and *pat.* (Beachman, of London, *inv.* self-acting water-closet and J. Dawson a portable water-closet.)

Water-frame (*See* Spinning, loom.)

Water-gas 1823. William de Vere and Henry S. Crane take out first *pat.* for illumination by water-gas. 1833. Jobard (Bel) *inv.* process for making water-gas from resins. (Process sold to Mme. Sellique and Tripier, of Paris and *pat.* in England and Austria by Jean de Marino.) 1847. Stephen White, of Manchester and H. M. Paine, of Worcester, Mass., U.S., *pat.* processes.

Water-gauge 1776. *Intro.* by James Watt instead of the previously used pett-cocks.

Water-glass 1644. Mixture mentioned by Johann Rudolf Glauber (1604–70). 1821. Von Fuchs *inv.* process of making. 1845. Ransome, of Ipswich *inv.* process. 1857. Kuhlmann, of Lille *inv.* process using water-glass for hardening stone of buildings, to preserve them.

Water-meter Running water first hydraulically measured by Castelli (1577–1644). 1824. Diaphragm type *inv.* 1856. Siemens *inv.* inferential type. Kennedy *inv.* piston type.

Waterproof Cloth 1815. Process for making *inv.* by James Syme (Scot) (1799–1870). 1835. Syme's process adopted and *pat.* by MacIntosh.

Waterwheel 13th cent. Crude sketch of undershot waterwheel appears in English MSS. 1405. Painting of overshot waterwheel preserved at Göttingen, Germany. 1556. Overshot waterwheel up to 40 ft. in diameter and giving up to 8 h.p. being *const.* 1582. Peter Morice *const.* tidal-driven waterwheel for London waterworks. 1588. A. Ramelli (1531–90) *desc.* under- and overshot waterwheels. 1682. S. Rannequin (Fr) installed waterwheel totalling over 100 h.p. at Marly Waterworks, France. (Some power was transmitted by connecting-rods over ¼ mile.) *c.* 1700. Waterwheel 40 ft. in diameter replaced one 12 ft. in diameter in Cornwall. *c.* 1750. Overshot waterwheel *inv.* (*sic*) by John Smeaton. 1826. Feathering waterwheel *inv.* by John Oldham. (The so-called "Whistling device" was the first evidence of the overshot

waterwheel, to which it led.) 1865. 72 ft. diameter, 150 h.p. waterwheel erected at Laxey, Isle of Man. 1880. Pelton waterwheel *inv.* in California. (*See* Turbine, water.) 1884. J. Duponchel (Fr) *inv.* twin, contra-rotating waterwheels. (*See also* Mills, water.)

WATSON, William (1834–1915) Pioneer of early experiments in cylinder and disc-type electrostatic machines.

WATT, James (1736–1819) 1763. *Des.* his first expansion steam-engine, to which he later made double-acting, and fitted with a centrifugal governor. His engines worked under what is termed "low-pressure." 1784. Formulated that 33,000 lb. raised 1 ft. equalled 1 h.p. 1794. *Inv.* the steam indicator. 1810. With Matthew Boulton *inv.* machines for the Royal Mint. (*See* individual items.)

Waves, Electromagnetic 1887. R. H. Hertz successfully *prod.* electromagnetic waves of 3 metres wavelength. 1887. Sir Oliver Lodge *prod.* electromagnetic waves a little earlier than Hertz and first sent Morse code signals by radio. Nikola Tesla first *prod.* long electromagnetic waves. (*See also* Radio telegraphy.)

Wax, Horsley's (surgery) *c.* 1885. *Inv.* by Sir Victor Horsley (1857–1916).

Wax, Sealing 1554. Oldest known letter sealed with wax. 1640. Mentioned by Francis Rosseau (Fr), of Auxerre. 1844. Finally superseded by adhesive envelopes.

Weather Telegraphy 1849. First tried by London *Daily*

News. 1859. First systematic weather forecasting *intro.* by Admiral Fitzroy. 1863. First international system *intro.*

Weather Scale 1805. Devised by Admiral Sir Francis Beaufort.

WEBER, Ernst Heinrich (Ger) (1795–1878) 1825. Pioneered study of acoustic wave-motion and interference.

WEBER, Wilhelm Édouard (Ger) (1804–91). 1825. Collaborated with his brother (above), and also pioneered work in electric telegraph design.

WEBSTER, Noah (1758–1843) 1783. Compiled *Spelling-book* of the English language. 1807. Compiled first U.S. dictionary of the English language, which was finished in 1825.

WEDGWOOD, Josiah (1730–95) 1759. Pioneered *prod.* of very hard pottery at his steam-engine-driven factory at Burslem.

Weedkillers, Selective *c.* 1945. "Methoxone" (4 chlor-2 methyl - phenoxyacetic acid) *disc.*

Weighing Machine 1821. Hydraulic weighing machine *inv.* by M. Henry.

Welding, Arc 1801. Electric arc first *prod.* by Sir Humphry Davy. 1880. Fusion of metal into a weld by Slavianoff (Rus), and by Charles Coffin, of Detroit, U.S. 1881. First installation at Milton, Staffordshire, for reduction of aluminium. First attempt at arc welding made by De Meritons. *c.* 1886. Arc welding process devised by Prof. Elihu Thompson. 1887. Carbon rods first used in arc welding by M. V. Bernardos (Rus).

Welding, Oxy-hydrogen 1801. Hydrogen-oxygen flame *inv.* by Robert Hare, of Philadelphia, U.S. 1847. Hare reduced 2 lb. of platinum in the oxy-hydrogen flame. 1895. Le Châtelier (Fr) measures heat of oxy-hydrogen flame. 1900. Edmund Fouché *dev.* first oxy-hydrogen torch.

Welding, Thermit *Inv.* by Goldschmidt.

Well, Artesian 1794. First in England sunk by Benjamin Vulliamy at Notting Hill, London. 1841. Mulot completes 1,798 ft. artesian well at Grenelle, near Paris (Artois).

WELSBACH, Carl Auer von (Ger) (1858–1929) 1897. Succeeded in *prod.* malleable tantalum. 1886. *Inv.* the gas-mantle.

WENZEL, Karl Friedrich (Ger) (1740–93) *c.* 1777. Pioneered study of chemical association and dissociation.

WESTINGHOUSE, George (U.S.) (1846–1914) 1880–90. Pioneered *des.* of electric transformers.

WHEATSTONE, Sir Charles (1802–75) Prolific inventor. 1837. *Inv.* with William Cooke, the needle telegraph, which was used between London and Slough, Buckinghamshire, in 1838. 1845. *Des.* first dynamo with an electrically energised field.

Wheel Pre-3000 B.C. Wooden disc wheels revolving on fixed axles in use in Sumeria. 2000–3000 B.C. Tripartite wheels of wooden planks in use. Post-2000 B.C. Wheels with leather or copper tyres and copper-studded rims in use. Spoked wheels appear on painted clay models and carved seals from Mesopotamia; being *intro.* into Egypt and Crete soon after 1600 B.C., some having six or eight spokes. 500 B.C. Celtic wainwrights of Bohemia and the Rhineland made spoked, dished wheels. Pliny (A.D. 23–79) mentions wheeled plough. A.D. 200. 15 ft., 22-spoked water-raising wheels *dev.* by Roman mining engineers at Rio Tinto, Spain. 1523. Fitsgerald mentions wheeled ploughs in England. 1805. Artillery wheels *inv.* by Samuel Miller. 1808. Wire-spoked, cycle-type wheel *inv.* by Sir George Cayley. (The tension wheel.) 1855. Keyed-on steel tyres *inv.* by Edward Turner for railway wheels.

Wheel, Autocycle 1897. Coventry Motor Company *intro.* ladies' friction-drive unit. 1900. Singer autocycle wheel *intro.* 1914. Wall auto-wheel *intro.* 1919. "Skootamota" *intro.* by A.B.C. Company. 1919. "Nera-Car" *des.* by C. A. Neracher. 1919. Lambretta-type, scooterbike *intro.* by Pullin. 1926. "Ro" monocar scooter *intro.*

Wheel-barrow *c* A.D. 230. Mentioned in Chinese literature. 12th cent. Mentioned in Western civilization. Mentioned by Blaise Pascal (Fr) (1623–62).

Wheel, Interchangeable motor-car 1904. "Stepney" spare rim *inv.* by T. M. and William Davies. 1905–06. Rudge-Whitworth detachable wheel *inv.* by J. V. Pugh. 1906. Detachable rim *inv.* by Louis Henry Perlman (U.S.). 1908.

Warland dual rim *inv.* by P. W. Turquand and S. H. Cope. 1908. Sankey interchangeable wheel of pressed steel *inv.* by Joseph Sankey. 1909. Captain detachable rim *inv.* by A. F. Gunstone.

Wheel, Prayer 400. Vertical type in common use in China. *c.* A.D. 1140. Yeh Meng-tê (*d.* 1148) mentions wind-powered temple prayer wheels. *See also* Governor, Centrifugal.

Wheel, Spring-spoked 1835. Iron-spoked "suspension" wheel *inv.* by Theodore Jones. Sprung wheel *inv.* by William Adams. 1875. Spring wheel *inv.* by De Mauni. 1902. sprung wheel *inv.* by Rousel Lecomte. 1904. "Glyda" sprung wheel with resilient hub *intro.* 1904. "Empire" spring-spoked wheel *intro.*

Whistle, Galton *Inv.* by Samuel Galton. (*See also* Siren.)

WHITNEY, Eli (U.S.) (1765–1825) 1793. *Inv.* the cotton gin, which was *imp.* three years later by Holmes.

WHITTLE, Sir Frank (*b.* 1907) *Des.* jet engine for aircraft which first flew in 1941 to provide practical experience and theoretical knowledge for the *dev.* of the gas-turbine. (*See* Aircraft.)

WHITWORTH, Sir Joseph *c.* 1850. Pioneered accurate toolmaking, producing a micrometer capable of reading to 1–10,000th of an inch. In 1844 produced a perfectly true plane surface-plate.

WIEN, Wilhelm (Ger) (1864–1928) 1893. *Prop.* the optical Law of Displacement.

WILCKE, Johan Carl (1732–

96) Pioneered researches in calorimetry, in terms of the substance theory of heat.

WILKINSON, John (1728–1808) Ironmaster who, in 1775, *inv.* an *imp.* cylinder-boring machine which he installed at his Bersham Ironworks. (*See also* Smeaton, John.)

WILDE, Henry 1863. *Inv.* his separately excited dynamo.

WILLANS, P. W. (1851–92) 1874. *Inv.* high-speed steam-engine. 1885. *Inv.* high-speed central-valve steam-engine.

WILLIAMS, Charles Greville (1829–1910) 1860. *Prod.* isoprene (artificial rubber).

WILLIS, Robert 1849. *Des.* one of the first traction-engines. It was *const.* by E. B. Wilson, of Ransomes and May.

WILLSLÄTTER, Richard (1872–1942) 1906–13. Investigated composition of plant pigments and *disc.* composition of chlorophyll.

Winds, Rotation of (Law of Storms) *c.* 1850. Law enunciated by Dove. (*See also* Isotherms.)

Winds, Scale of 1806. Devised by Admiral Sir Francis Beaufort.

Windmill Legendary: First windmill (having eight sails) *const.* for Caliph Omar by his Arab slave Abū Lutuā. 1170. Windmill mentioned in a charter as existing at Swineshead, Lincolnshire. 1180. Post type windmill mentioned by Léopold de Lisle as existing in Normandy. 1185. Windmill at Weedley, Yorkshire valued at 8s. per year. 1189. Windmill erected at Dunwich, Suffolk. 1191. Windmill built at Bury St. Edmunds. *c.* 1270 Earliest *illus.* of windmill in England *Windmill Psalter*, probably written at Canterbury. 1344. Documentary evidence of use of windmills for drainage in Holland. 1420. First *illus.* of tower windmill appears in French MSS. 1430. Water-draining scoop-wheels *intro.* into Holland at Polder. 1438–50. *Illus.* of first vertical axle windmill appears in un*pub.* notebook of Mariano Jacopo Taccola. 15th cent. First *illus.* of tower windmill in England appeared. 16th cent. Windmill sails with varying angle of incidence first used. 1589. Cornelis Dircksz Muys (Hol) *inv.* mud-pumping "tjasker," windmill-driven. Muys (Hol) *inv.* double-curved sails, or "zeeg." 1593. Cornelis Cornelisz (Hol) *inv.* first windmill-driven sawmill, and in 1597 a windmill-driven edge-runner mill for crushing oil seeds. 1745. Edmund Lee *inv.* automatic fantail wind-facing device. 1750. Andrew Meikle (1719–1811) *inv.* self-reefing sails. Bevel-wheels *intro.* into windmill machinery. 1754. John Smeaton *intro.* cast-iron into windmill *const.* 1759. Smeaton experiments on model windmills to find best proportions. 1769. Smeaton *intro.* use of cast-iron for windmill gearing. 1772. Andrew Meikle *inv.* spring-sail with multiple hinged shutters. 1787. "Lift-tenter," with centrifugal governor *inv.* by Thomas Mead. 1789. Stephen Hooper *inv.* roller-reefing sail. 1807. William Cubitt (1795–1861) combined

Meikle's and Hooper's *invs.* in his self-reefing sails. 1810. Water-pumping windmill erected in Scotland. 1826. Horizontal, drum-vaned windmill appeared. 1836. Horizontal windmill *pat.* by William Symington. (*See also* Panemour) 1840. "Berton" windmill sail *inv.* 1860. Air-brake *inv.* by Catchpole, of Sudbury. 1924. Dekker (Hol) *intro.* aerofoil-section sails and roller shaft bearings, giving a 300 per cent increase in power output. 1941. Windmill electric generator with 200 ft. diameter, two-bladed sails erected at Grandpa's Knob, Rutland, Virginia, U.S. (This was the first wind-powered public electricity supply in the world.) 1955. 100-kilowatt, hollow-bladed windmill driving an air turbine erected at St. Albans, Hertfordshire.

Windscreen Wiper (motor-car) 1920. Suction type *inv.* by W. M. Folberth. 1923. Electric type appeared in U.S. (1925, in England).

Wind-tunnel 1871. First made and used by Francis Herbert Wenham and John Browning *c.* 1880. H. F. Phillips made wind-tunnel 6 ft. long and 17 in. square with air-flow *prod.* by a steam injector. This was first use of wind-tunnel to determine lift and drag of plane and cambered vanes.

Window-winders, Motor-car 1930. Geared type *inv.* by H. M. Hobson.

Windows, Glass 650 *Intro.* by Bishop Biscop in English churches. 1177. First built into English private houses.

Winnowing Machine 1710. *Intro.* into England from Holland. *See also* Grain-cleaning machines.

Wire Wire *prod.* for filigree work in Ur of the Chaldees by cutting a continuous strip from a circular metal sheet. 2500 B.C. Wire *prod.* by drawing through dies in Egypt. 10th cent. Draw-plates in use at Augsburg. 11th cent. Brass wire mentioned. *c.* 1400 Wire *inv.* (*sic*) by Rudolph of Nuremburg. 1491. Wire-drawing machine mentioned by Conrad Celtes as having been *inv.* by Rudolph of Nuremburg. 1565. Wire *intro.* into England by the Dutch. 1663. First wire drawn in England at Mortlake, Surrey.

WISLICENUS, Johannes (1835–1902) 1872. *Disc.* two types of lactic acid, only one of which was optically active: the principle of molecular asymmetry.

WÖHLER, Friedrich (Ger) (1800–82) 1827. *Disc.* beryllium. 1827. *Disc.* aluminium.

WOLLASTON, Dr. William Hyde (1766–1828) 1803. *Disc.* palladium, and also *inv.* the camera lucida.

Wood-joints 3000 B.C. In use in Ancient Egypt.

Wood Preservation 1779. May, of Amsterdam, impregnates wood with preservative. 1794. Sir Samuel Bentham experiments with vacuum impregnation. (*See also* Creosoting and Kyanizing.)

Wood's (Fusible) Metal 1860. Alloy of cadmium, tin, lead, and bismuth *pat.* by Dr. B. Wood, of Nashville, U.S.

Woodworking Machinery

1793. Sir Samuel Bentham *inv.* woodworking machinery for planing, rebating, mortising and curved and transverse sawing, to make complete window-sashes. 1802. Joseph Bramah *inv.* first rotary wood-planer.

WORBY, William 1842. *Des.* steam tractor built by Ransomes and May and exhibited at Bristol Royal Agricultural Show.

WORCESTER, Marquis of (1601–67) 1663. *Des.* primitive steam pumping-engine. Wrote book *Century of Inventions* which *desc.* 100 novel ideas, some of which were not brought into use for many hundreds of years.

WORTHINGTON, Henry Rossiter (1817–80) 1841. *Inv.* the direct-action steam pump.

WRIGHT, Thomas 1845. Took out first British *pat.* for electric arc-lamp. (Clockwork-rotated, carbon discs.)

WRIGHT, Brothers Wilbur and Orville (U.S.) 1900. Commenced gliding experiments at Kill Devil Hills. 1902. Glided 200 yards. 1903. Made first powered aeroplane flight with a 16 h.p. engine mounted on a glider. 1905. Made powered flight of $24\frac{1}{4}$ miles. (*See* Aircraft.)

Writing 4000 B.C. Pictograms with 2,000 signs used at Erech, Sumeria. 3100 B.C. Heiroglyphics first appear in Egypt on King Narmer's Palette. 3000 B.C. Cuneiform writing with shaped reed-end on clay. 2900 B.C. 2,000-charactered Sumerian writing reduced to 500–600 signs, of which 100 represented syllables. 1594 B.C. Writing first taught to Latins by Princess of Phoenicia.

WROBLEWSKY, Zygmunt Florenty von (1845–88) *c.* 1883. One of the pioneers of cryogenic physics.

Xanthian Marbles 1845. *Disc.* by Sir Charles Fellows.

Xenon (element) 1898. *Disc.*

Xerography 1937. *Inv.* by Chester Carlson (U.S.).

X-rays 1895 (Nov.). *Disc.* by Wilhelm Conrad Röentgen (1845–1923), of Wurzburg.

(*Disc.* announced Jan. 23, 1896.)

X-ray Screen (Barium platino-cyanide) 1853. Effect *disc.* by Stokes.

X-ray Tube, High-voltage 1913. *Inv.* by Dr. Coolidge (U.S.).

Yard 1305. Standardized by King Henry I of England as, the length of his arm.

Year, Length of the Determined to withn 12 seconds of accuracy by Hipparchus (160–145 B.C.). (Tropical year.)

YOUNG, Thomas (1773– 1829) Important contributor to 19th cent. science of acoustics and optics.

Ytterbium (element) 1878. *Disc.* by Jean Charles Glissard de Marignac (Swit).

Yttrium (element) 1794. *Disc.* by Johann Gadolin (1760– 1852).

Z

ZAMBONI, Guiseppe (It) (1776–1846) 1810. Re-*disc.* the dry pile electric battery of Georg Bernhard Behrens (1802).

Z-crank Engine 1855. *Inv.* by Morton and Hunt, of Glasgow.

ZEEMAN, Pieter (Hol) 1896. With H. A. Lorentz (Hol) *disc.* the connection between light and magnetism—the Zeeman effect.

Zeeman Effect 1885. *Obs.* by Charles Fieviez, but ignored by him. 1896. Pieter Zeeman *disc.* connection between light and magnetism. 1897 J. J. Thompson proved electron charge at least 1,000 times

as large as that of hydrogen atoms by study of the "Zeeman effect." 1915. A. Semmerfeld (Ger) applied relativity to the Zeeman effect. 1919. J. Starke (Ger) used electricity to produce the effect, where Zeeman had used magnetism.

ZEISS, Carl (1816–88) One of the pioneers of scientific optical work and lens-making.

Zero (symbol) *Inv.* in India in the early A.D. cents.

Zero, Absolute 1908. Kammerlingh Omnes *prod.* a temperature of 4.2° above absolute zero. 1926. Debye and Giauque suggest use of electro-magnetic field to *prod.* absolute zero. 1933. Giauque and De Hass reduced temperature of gadolinium sulphate to 1° above absolute zero. 1939. Ashmead, of Cambridge, reduced temperature of ammonia-copper sulphate to 0.002° above absolute zero.

"Z.E.T.A." (Zero Energy Thermo-nuclear Assembly) 1958. Announced as being in use.

Zinc *c.* 500 B.C. Zinc found in hollow silver bracelets found at Rhodes. 1231. Zinc ore first noticed by Europeans. Zinc first mentioned as a distinct metal by Paracelsus (Swit) (1493–1541). 1721. Zinc *isol.* in Western world by Henckel. 1738. *Intro.* to Western industry by Champion. 1809. Zinc mines *disc.* near Craven, Yorkshire.

Zincography 1815. *Intro.* into England.

Zirconium (element) 1786.

Disc. by M. H. Klaproth. 1792. Independently *disc.* by Vauquelin (Fr). 1914. First obtained as a pure metal.

Zöetrope 1824 (? 1833). *Inv.* by three experimenters.

Zoology Science founded by Conrad Gesner (Swit) (1516–65), of Zurich. He was the father of modern zoology and wrote *Historiae animalum.*

ZSIGMONDY, Richard Adolf (Ger) (1865–1929) 1903. *Inv.,* with Siedentopf, the ultramicroscope.

ZUCCI, Niccolo (It) (1586–1670) Planned and attempted to *const.* a reflecting telescope.

Zurich Machine (rotary water-pump) 1746. *Inv.* by Wirtz, a dyer of Limmat, near Zurich.

ZWORYKIN, Vladimir K. (U.S.) *c.* 1934. *Inv.* the iconoscope television camera. (*See* Television.)

NOTES